Wireless Networks

Series Editor

Xuemin Sherman Shen, University of Waterloo, Waterloo, ON, Canada

The purpose of Springer's Wireless Networks book series is to establish the state of the art and set the course for future research and development in wireless communication networks. The scope of this series includes not only all aspects of wireless networks (including cellular networks, WiFi, sensor networks, and vehicular networks), but related areas such as cloud computing and big data. The series serves as a central source of references for wireless networks research and development. It aims to publish thorough and cohesive overviews on specific topics in wireless networks, as well as works that are larger in scope than survey articles and that contain more detailed background information. The series also provides coverage of advanced and timely topics worthy of monographs, contributed volumes, textbooks and handbooks.

** Indexing: Wireless Networks is indexed in EBSCO databases and DPLB **

Xiang Cheng • Shijian Gao • Liuqing Yang

mmWave Massive MIMO Vehicular Communications

Springer

Xiang Cheng
Peking University
Beijing, China

Shijian Gao
University of Minnesota
Minneapolis, MN, USA

Liuqing Yang
The Hong Kong University of Science and Technology
Guangzhou, Guangdong, China

ISSN 2366-1186 ISSN 2366-1445 (electronic)
Wireless Networks
ISBN 978-3-030-97510-4 ISBN 978-3-030-97508-1 (eBook)
https://doi.org/10.1007/978-3-030-97508-1

This Springer imprint is published by the registered company Springer Nature Switzerland AG
The registered company address is: Gewerbestrasse 11, 6330 Cham, Switzerland

Preface

The automobile industry is currently shifting from "driving by humans" to "driving by intelligence." Such a transformative evolution is primarily propelled by fast-increasing onboard sensors, with which the vehicles are gaining unprecedented degrees of intelligence. Numerous sensors will inevitably bring in massive sensory data, making reliable and swift information transfer an urgent and crucial issue. Existing wireless solutions include DSRC and LTE-V2X, both permitting low-volume message delivery. However, due to the limited bandwidth, they are still far from meeting the Gbps-level data rate regulated by the automobile industry. The current limitation inevitably motivated the exploration of the mmWave band, where vast spectrum resources are available for high-speed information transfer. Besides the bandwidth merit, mmWave's inherent short wavelength allows a natural combination with massive MIMO for higher diversity and multiplexing gain. In theory, alternating the operating frequency does not need to change the wireless regime, but the fact is that many implementing concerns, such as power consumption and hardware expenditure, prohibit mmWave systems from inheriting classic fully-digital transceivers. Instead, an economical yet restricted structure, namely the hybrid beamformer, comes into practical use. In conjunction with the higher signal dimensions and complicated channel environments, the compromised hardware architecture requires a paradigm-shifting design to underpin mmWave communication. Against this background, this book will showcase a comprehensive picture regarding advanced mmWave massive MIMO techniques, hoping to provide promising physical-layer solutions to vehicular communications in the 5G and Beyond era.

The book is organized as follows. Chapter 1 overviews vehicular communications and elaborates the necessity of mmWave technologies. Chapter 2 introduces state-of-the-art mmWave channel modeling, with space-time-frequency and non-stationary features taken into account. Based on the insights from channel modeling, Chap. 3 presents an efficient channel estimator dedicated to mmWave transceivers with hybrid structures, which is capable of combating doubly selective massive MIMO mmWave channels. The obtained channel state information opens the door to a generic mmWave multi-user transceiver design, with the detailed strategies presented in Chap. 4. Driven by the pursuit of lower error rate and higher energy efficiency, Chaps. 5 and 6 explore the potential use of index modulation in hybrid mmWave systems. Although both chapters will deal with the doubly selective channels, Chap. 5 focuses on the uplink multi-user access, whereas Chap. 6 spotlights on downlink multi-user transmission. Despite our best effort, the above content covers just a tip of the iceberg of mmWave vehicular communication. Thus, some open problems and promising directions will be discussed at the end of each chapter for future studies.

This book is mainly oriented to researchers, graduated students, and professors relevant to this field. Nevertheless, it also serves as a great introduction to state-of-the-art mmWave vehicular communications for those outside this field but aspire to pursue new interdisciplinary directions.

We would like to thank Ms. Yajun Fan, Mr. Ziwei Huang, and Mr. Zonghui Yang for their inspiring discussions on the research work presented in this book. Finally, we would like to thank the continued support from the National Natural Science Foundation of China under Grant 62125101 and the National Science Foundation under Grants ECCS-2102312 and CNS-2103256.

Beijing, China — Xiang Cheng
Minneapolis, MN, USA — Shijian Gao
Guangzhou, China — Liuqing Yang

Contents

Acronyms

3GPP	The 3rd Generation Partnership Project
ADC	Analog-to-Digital Converter
APEP	Average Pairwise Error Probability
APS	Analog Phase Shifter
BER	Bit Error Rate
BS	Base Station
CDL	Clustered Delay Line
CP	Cyclic Prefix
CS	Compressed Sensing
D2D	Device-to-Device
DAC	Digital-to-Analog Converter
DSRC	Dedicated Short-Range Communications
FFT	Fast Fourier Transform
GMD	Geometric Mean Decomposition
GPS	Global Positioning System
IFFT	Inverse Fast Fourier Transform
IM	Index Modulation
ITS	Intelligent Transportation System
IV	Internet of Vehicles
LTE	Long-Term Evolution
MIMO	Multiple-Input Multiple-Output
MS	Mobile Station
NMSE	Normalized Mean Square Error
NR	New Radio
OFDM	Orthogonal Frequency Division Multiplexing
OMP	Orthogonal Matching Pursuit
PSD	Power Spectrum Density
QoS	Quality of Service
RF	Radio Frequency
RSU	Roadside Unit
SM	Spatial Modulation

SNR	Signal-to-Noise Ratio
SVD	Singular Value Decomposition
TDL	Tapped Delayed Line
UE	User Equipment
V2V	Vehicle-to-Vehicle
V2X	Vehicle-to-Everything
VANET	Vehicular Ad hoc Network
VCN	Vehicular Communication Network

Chapter 1
Millimeter-Wave Vehicular Communications

Abstract This chapter works on presenting a background overview associated with vehicular communications and networking. The first part focuses on reviewing the recent advancements and progress booming in vehicular networking and standardization. As existing wireless solutions cannot guarantee safe, swift, and ubiquitous high-volume data transfer in future vehicular-to-everything communication, mmWave comes into play to help address the information bottleneck beyond the 5G era. Henceforth, the second part follows to provide a holistic overview of mmWave physical fundamentals and mmWave system properties. The chapter is concluded with a brief organization of this monograph studying mmWave vehicular communications.

Keywords Vehicular communication · Networking · mmWave · 5G · Standardization · Vehicular-to-everything

1.1 Overview of Vehicular Communications

The idea of intelligent vehicles (IV) was proposed more than three decades ago [1]. In the nascent stage of the relevant studies regarding IV, or its generalized form, intelligent transportation systems (ITS), the focus has been mostly on the conceptual aspects [2–4]. Thanks to the recent evolution booming in the automotive industry, IV is attracting increasing attention. From the perspective of vehicle itself, more sensors and antennas are now being mounted to augment its sensing and communication capabilities [5]. From the infrastructure aspect, numerous roadside units and cloud-based networks are now being deployed to underpin ubiquitous information exchange and storage [6, 7]. From the technological side, advanced communication regimes and deep learning methods play a crucial role in information transmission and interpretation [8]. From the administration level, scientific regulation and policy-making help to cope with the potential risks and boost the commercialization. It is envisioned that completely self-driving vehicles will be ready for practical deployment in the next decade (see the roadmap in Fig. 1.1).

X. Cheng et al., *mmWave Massive MIMO Vehicular Communications*,
Wireless Networks, https://doi.org/10.1007/978-3-030-97508-1_1

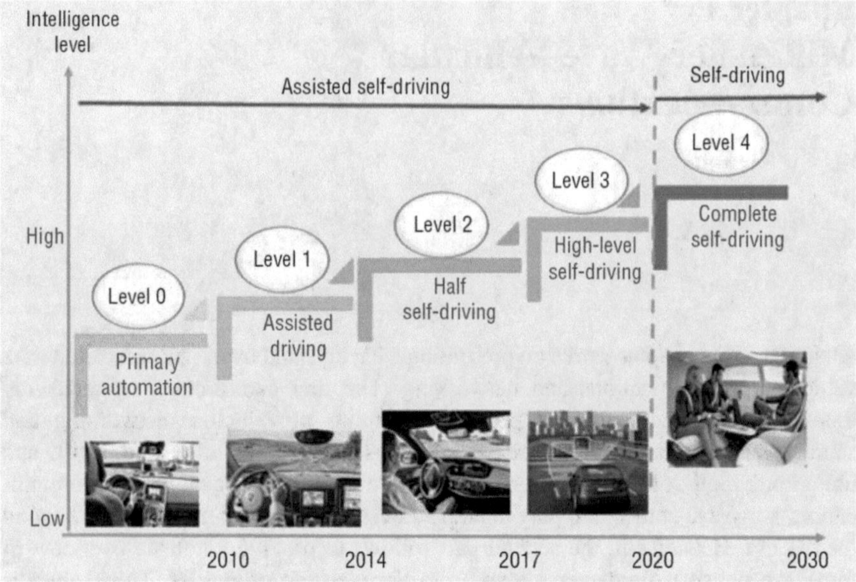

Fig. 1.1 The evolutionary roadmap of autonomous driving [9]

Although the success of IV relies on an array of technologies, vehicular communication and networking (VCN) still comes as a fundamental one. Such a claim is of no exaggeration. If one focuses on a single vehicle, its primary source of intelligence comes from parsing the sensing information, either via onboard sensors or shared by other vehicles. With hundreds of thousands of cars on the road, the amount of information flowing into the network would be astronomical. The data deluge, accompanied by many other challenges, such as driving safety, transportation efficiency, reaction promptness, as well as decision reliability, has posed a significant burden on the information tunnel, making the role of VCN much more crucial than ever [9].

The most representative type of VCN was rooted from the ad-hoc network architecture in the mobile environment, termed as vehicular ad-hoc network (VANET) [10, 11]. Both the industry and the academia have devoted considerable efforts to standardize VANET. The primary outcome is the well-known dedicated short-range communications (DSRC) standard. Although DSRC and its variations have been adopted by some major economic entities, widespread deployment still has a long way to go because the infrastructure expenditure is prohibitive. In fact, the financial issue is not the only bottleneck. Recall that DSRC stemmed from IEEE 802.11p. The latter exhibits weakness in scalability and mediocre QoS as the network size grows, limiting the performance in data rate and throughput. As a result, VANET can hardly satisfy the demanding vehicular communications needs.

In light of the deficiencies of VANET, another direction of standardizing VCN is to seek the power of cellular networks. Compared with VANET, cellular

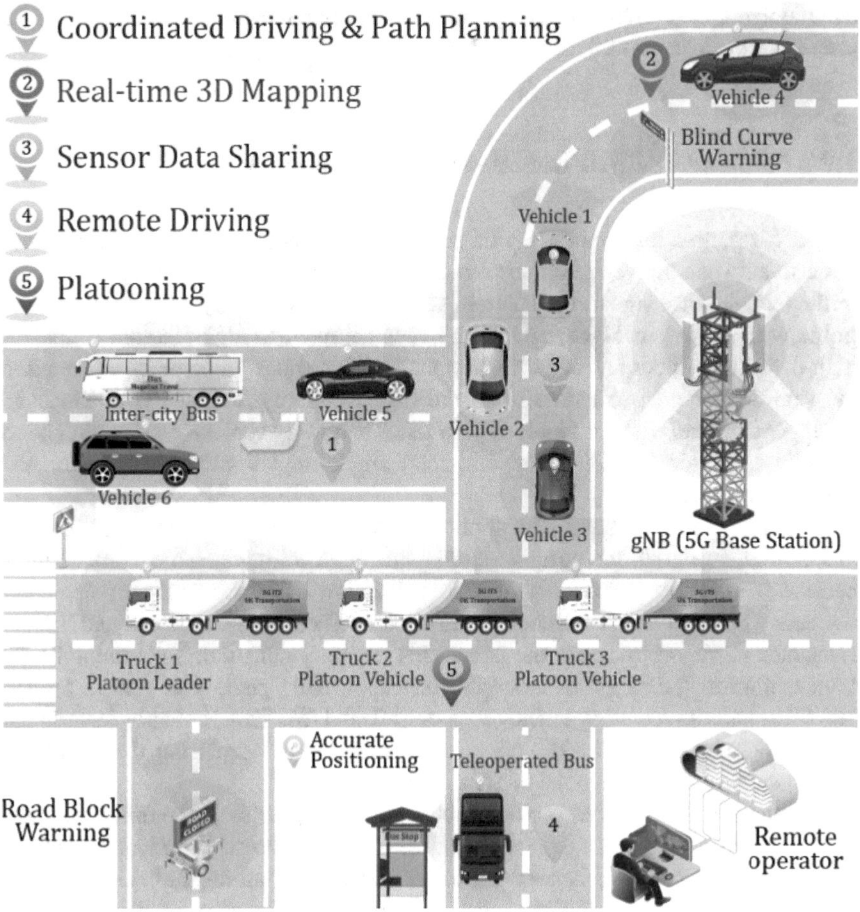

Fig. 1.2 Advanced use cases and services envisioned in 5G-V2X [12]

networks are undoubtedly far more mature in development, standardization, and commercialization. Over the past half century, cellular systems have evolved from the primitive 1G analog communication to today's 5G new radio (NR) that enables secure seamless wireless connections over a wide spatio-temporal span. The combination of state-of-the-art 5G cellular networks and vehicle-to-everything (V2X) communications, namely 5G-V2X, has been recognized as an overall enhanced alternative to VANET. Some promising use cases and services in 5G-V2X are shown in Fig. 1.2. There are three primary reasons that 5G-V2X is expected to become the new mainstream. First, it leverages the existing cellular infrastructures instead of building from scratch. Secondly, its centralized architecture makes it more capable in network control and traffic prioritization. Thirdly, it includes a vital feature termed as the proximity service (ProSe) that stems from device-to-device (D2D) communications. Hence it can replace DSRC to support vehicle-to-vehicle (V2V) communications with low latency [13]. These prominent benefits render 5G-V2X a promising business model for future ITS.

1.2 Necessity of Millimeter-Wave Technology

In VCN, frequent information exchanges will occur not only among the vehicles but also between the vehicles and infrastructures [14]. Apart from the low-volume traffic (e.g., control signaling, location, velocity), many other types of high-volume traffic will also be involved, including but not limited to high-resolution images, high-definition videos and 3D localized maps. These huge amounts of information must be delivered in real time to ensure high reliability in decision making and intelligent control for autonomous driving. Therefore, vehicular communication urgently calls for multi-Gigabit data rate, single-digital millisecond level, and ubiquitous connection.

Although numerous communication techniques have been proposed to approach this challenging goal, the current state is still far from being satisfactory. This is because the operating frequency of cellular is below 6 GHz, and the corresponding frequency band co-exists with many other bands. The overly congested and limited frequency cartography limits the achievable capacity potential. Take the 4G long term evolution (LTE) as an example. Its supportable peak data rate is around 100 MHz and the resulting latency is about 50 ms [15]. This is clearly insufficient for VCN. Thus scaling up the operating frequency and exploring the uncharted frequency band become a must in 5G & Beyond era.

Fortunately, 5G NR does provide such an option by utilizing the mmWave band that ranges from 30 to 300 GHz (see Fig. 1.3). Compared to the prevalent sub-6 GHz band, the mmWave band is over ten times wider. Such an unparalleled spectrum abundance effectively facilitates intensive information transfer. In the meantime, the maturity in mmWave device fabrication also lays a solid cornerstone for the fast roll-out of 5G cellular. Since mmWave technologies can be treated as an adds-on benefit without interfering the sub-6 GHz cellular much, it can effectively mitigate the information bottleneck in today's VCN [17].

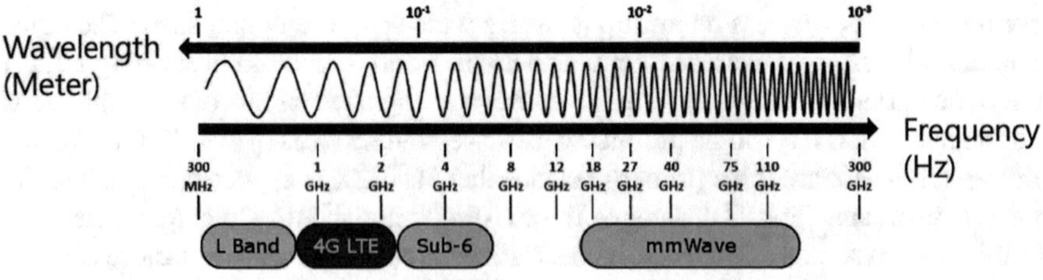

Fig. 1.3 The networking spectrum bands [16]

1.3 Characteristics of Millimeter-Wave Systems

Same as current sub-6 GHz waves, mmWave also belongs to the set of electromagnetic waves except for having a much shorter wavelength. However, it is this unique physical property that leads to a distinct system design [18]. Specifically, mmWave's higher frequency induces higher propagation attenuation, rendering it challenging for a single antenna to serve long-distance communications. It is well known that massive antennas can help overcome the propagation loss thanks to their high array gains. Fortunately, mmWave's shorter wavelength makes it natural to combine with massive multiple-input multiple-output (MIMO) [19, 20]. In conventional MIMO systems, each antenna is equipped with a complete radio-frequency (RF) chain. Therefore, digital signal processing can adjust both the symbol's amplitude and phase. In this sense, classic fully-digital structures are essentially free from any hardware constraints. However, the digital structure is not a practical solution to mmWave massive MIMO (mMIMO). This is because mmWave RF chains are much more expensive and power-consuming than their sub-6 GHz counterparts. To lower the implementation cost, mmWave mMIMO will typically adopt a more efficient structure, namely the hybrid structure [21]. Such a low-cost alternative consists of a large-scale yet economic analog network and much fewer RF chains.

Within the hybrid structure, the precoding comprises two stages: the low-dimensional digital processing and the high-dimensional analog processing. The latter is constrained in the sense that it can vary the symbol's phase only. This limitation, whist significantly lowering the expenditure, brings in many constraints and special considerations in the physical-layer design. However, a rich literature has shown that, with dedicated handling, this structure is able to yield a decent balance over cost and performance. Considering that, we will treat the hybrid structure as a default property of mmWave systems throughout this monograph.

1.4 Organization of the Monograph

The monograph is organized as follows. This chapter overviews vehicular communications and explains the necessity of mmWave technologies. Chapter 2 introduces state-of-the-art mmWave channel modeling, with space-time-frequency characteristics and non-stationary features taken into account. Based on the insights from channel modeling, Chap. 3 presents an efficient channel estimator dedicated to hybrid structures and capable of combating double selectivity. The obtained channel state information opens the door to a generic mmWave multi-user transceiver design, with the detailed discussions provided by Chap. 4. Driven by the pursuit of a lower error rate and a higher energy efficiency, Chaps. 5 and 6 explore the potential use of index modulation to hybrid mmWave systems. While the channel's double selectivity has been incorporated in both, Chap. 5 focuses on the uplink multi-user access, whereas Chap. 6 spotlights on downlink multi-user transmission. Despite our best effort, the above content covers just a tip of the iceberg of mmWave vehicular communications. Thus some open problems and promising directions are discussed at the end of each chapter.

References

1. X. Cheng, R. Zhang, L. Yang, Wireless toward the era of intelligent vehicles. IEEE Int. Things J. **6**(1), 188–202 (2019)
2. X. Cheng, D. Duan, L. Yang, N. Zheng, Societal intelligence for safer and smarter transportation. IEEE Int. Things J. **8**(11), 9109–9121 (2021)
3. G. Dimitrakopoulos, P. Demestichas, Intelligent transportation systems. IEEE Vehic. Technol. Mag. **5**(1), 77–84 (2010)
4. X. Cheng, X. Hu, L. Yang, I. Husain, K. Inoue, P. Krein, R. Lefevre, Y. Li, H. Nishi, J.G. Taiber, F.-Y. Wang, Y. Zha, W. Gao, Z. Li, Electrified vehicles and the smart grid: the ITS perspective. IEEE Trans. Intell. Transp. Syst. **15**(4), 1388–1404 (2014)
5. J. Choi, V. Va, N. Gonzalez-Prelcic, R. Daniels, C.R. Bhat, R.W. Heath, Millimeter-wave vehicular communication to support massive automotive sensing. IEEE Commun. Mag. **5412**, 160–167 (2016)
6. S. Fuller, G. Waters, *Smarter Infrastructure for a Smarter World* (2020). https://www.nxp.com/docs/en/white-paper/SMRTRINFRASTRWP.pdf
7. A. Ali, N. Gonzalez-Prelcic, R.W. Heath, A. Ghosh, Leveraging sensing at the infrastructure for mmwave communication. IEEE Commun. Mag. **58**(7), 84–89 (2020)
8. A. Ferdowsi, U. Challita, W. Saad, Deep learning for reliable mobile edge analytics in intelligent transportation systems: an overview. IEEE Vehic. Technol. Mag. **14**(1), 62–70 (2019)
9. X. Cheng, R. Zhang, L. Yang, *5G-Enabled Vehicular Communications and Networking* (Springer, Cham, 2018)
10. M. Booysen, S. Zeadally, G.-J. van Rooyen, Survey of media access control protocols for vehicular ad hoc networks. IET Commun. **5**(11), 1619–1631 (2011)
11. S.-I. Sou, O.K. Tonguz, Enhancing VANET connectivity through roadside units on highways. IEEE Trans. Vehic. Technol. **60**(8), 3586–3602 (2011)
12. H. Bagheri et al., 5G NR-V2X: toward connected and cooperative autonomous driving. IEEE Commun. Stand. Mag. **5**(1), 48–54 (2021)
13. X. Cheng, L. Yang, X. Shen, D2D for intelligent transportation systems: a feasibility study. IEEE Trans. Intell. Transp. Syst. **16**(4), 1784–1793 (2015)
14. S. Chen, J. Hu, Y. Shi, Y. Peng, J. Fang, R. Zhao, L. Zhao, Vehicle-to-everything (v2x) services supported by LTE-based systems and 5G. IEEE Commun. Stand. Mag. **1**(2), 70–76 (2017)
15. S. Chen, J. Hu, Y. Shi, L. Zhao, LTE-V: A TD-LTE-based V2X solution for future vehicular network. IEEE Int. Things J. **3**(6), 997–1005 (2016)
16. R. Triggs,What is 5G, and what can we expect from it? (2021). https://www.androidauthority.com/what-is-5g-explained-944868/
17. S. Chen, J. Hu, Y. Shi, L. Zhao, W. Li, A vision of C-V2X: Technologies, field testing, and challenges with Chinese development. IEEE Int. Things J. **7**(5), 3872–3881 (2020)
18. J. Brady, N. Behdad, A.M. Sayeed, Beamspace MIMO for millimeter-wave communications: system architecture, modeling, analysis, and measurements. IEEE Trans. Antennas and Propag. **61**(7), 3814–3827 (2013)
19. A.L. Swindlehurst, E. Ayanoglu, P. Heydari, F. Capolino, Millimeter-wave massive MIMO: the next wireless revolution? IEEE Commun. Mag. **52**(9), 56–62 (2014)
20. S. Han, I. Chih-Lin, Z. Xu, C. Rowell, Large-scale antenna systems with hybrid analog and digital beamforming for millimeter wave 5G. IEEE Commun. Mag. **53**(1), 186–194 (2015)
21. F. Sohrabi, W. Yu, Hybrid digital and analog beamforming design for large-scale antenna arrays. IEEE J. Select. Top. Signal Process. **10**(3), 501–513 (2016)

Chapter 2
Millimeter-Wave Massive MIMO Vehicular Channel Modeling

Abstract This chapter works on developing a novel three-dimensional (3D) non-stationary irregular-shaped geometry-based stochastic model (IS-GBSM) for beyond 5G and 6G vehicle-to-vehicle (V2V) mmWave massive multiple-input multiple-output (MIMO) channels. The proposed IS-GBSM utilizes distinguishable dynamic clusters and static clusters to explore the impact of vehicular traffic density (VTD) on channel statistics. Specifically, the developed method generates dynamic/static correlated clusters by an improved K-Means clustering algorithm. Then, by employing a birth-death process based on correlated groups, the consistency in birth and death between dynamic/static correlated clusters during time-array evolution is modeled. Finally, extensive simulations are carried out and demonstrate that space-time-frequency non-stationarity has been accurately captured, and the influence of VTDs on channel statistics has been successfully explored.

Keywords Geometry-based · Stochastic · Channel model · mmWave · Massive multiple-input multiple-output · Vehicle-to-vehicle · Vehicular traffic density · Space-time-frequency correlation

2.1 Introduction of Vehicular Channel Model

2.1.1 Vehicular Channel Characteristics

Compared with traditional cellular communications, the vehicle-to-vehicle (V2V) communication has several unique characteristics, e.g., (1) vehicles at transmitter (Tx) and receiver (Rx) are equipped with low-elevation antennas, and (2) Tx, Rx, and some scatterers/clusters in the environment are moving at high speed. Figure 2.1 shows a simple and typical vehicular communication scenario with dynamic vehicles and static roadside environments. The waves at the Tx pass through different obstacles in the environment and reach the Rx. In order to reflect a more practical V2V communication channel, it is necessary to distinguish between moving obstacles and static obstacles in the communication environment

X. Cheng et al., *mmWave Massive MIMO Vehicular Communications*,
Wireless Networks, https://doi.org/10.1007/978-3-030-97508-1_2

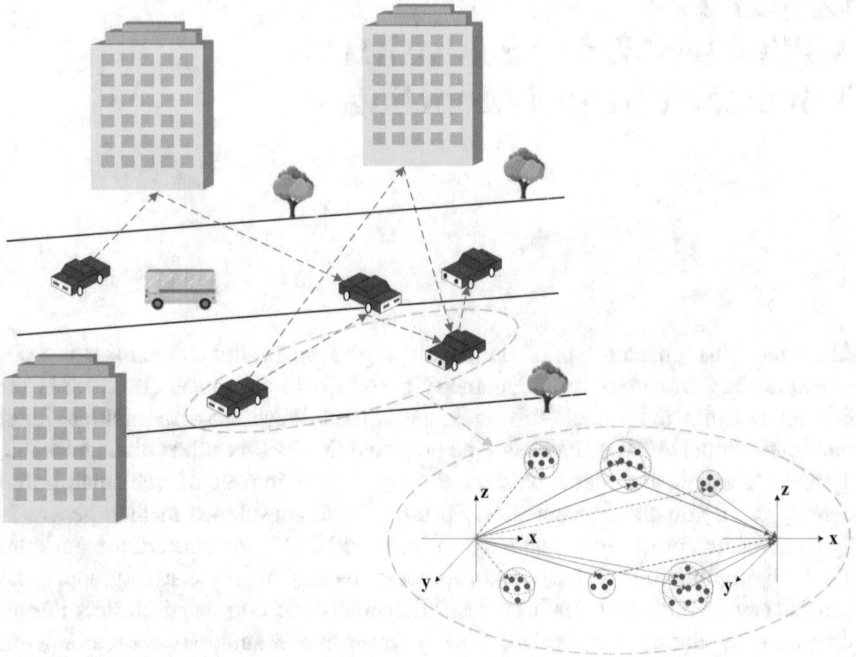

Fig. 2.1 A common vehicular communication scenario

[1]. At the same time, the influence of vehicular traffic density (VTD) on channel characteristics needs to be considered in the modeling procedure [2], such as rural street scenarios with low VTDs and fewer buildings, while urban street scenarios with more cars and buildings [3]. In addition, the existence of high-mobility moving vehicles will lead to changes in the surrounding communication environment and clusters. The rapid time-variation of channels results in the fact that the channel stationary interval will be shorter than the observation time. In such a case, the statistical properties of vehicular channels will fluctuate significantly over time [4, 5]. The fast time-variant property in the V2V communication environment results in the channel non-stationarity in the time domain, i.e., time non-stationarity. Note that channel non-stationarity in a certain domain means that the channel statistical characteristics vary in this domain. In this case, the V2V channel statistical properties are time-variant essentially [6].

2.1.2 Recent Vehicular Channel Model

It is well known that channel modeling is an enabling foundation for successful design of any communication systems [7, 8]. Considering the unique V2V channel characteristics, extensive work related to V2V channel modeling has been carried out, and the corresponding channel models have been proposed [9]. In light of the modeling approach, existing vehicular models can be categorized into geometry-based deterministic models (GBDMs) and stochastic models, and the latter can be further divided into non-geometry-based stochastic models (NGSMs) and geometry-based stochastic models (GBSMs). These diverse vehicular channel models will be elaborated in the sequel.

The GBDM characterizes vehicular physical channel parameters in a completely deterministic manner. The authors in [10] developed a V2V channel model based on the GBDM, where the ray-tracing technique is extended to reproduce the underlying physical radio propagation process for a given environment. Also, based on the ray-tracing technique, the V2V deterministic channel models [11, 12] can obtain the underlying characteristics of V2V channels accurately. However, the GBDM requires a detailed and time-consuming description of the site-specific propagation environment. As a consequence, the GBDMs in [10–12] are solely suitable for a specific propagation scenario and are not suitable for various scenarios, which may occur in realistic modeling.

For the NGSMs, the parameter of channels is determined in a stochastic way. In the NGSM, the representation of the wireless channel is stochastic and does not assume any scattering geometry. For the tap-based NGSM, its CIR is based on tapped delay line (TDL), and it relies completely on the statistical characteristic and probability density distribution (PDF) of the empirical parameter. Three types of V2V NGSMs developed in [13, 14] were more general and less complex compared with the GBDMs. Nevertheless, these three NGSMs in [14] for V2V channels ignored the non-stationarity in the time domain. To capture the time non-stationarity, the authors in [2] proposed a TDL-based NGSM, where the birth-death process method was developed to model the appearance and disappearance of multi-path components. With the increase of communication bandwidth, the measurement indicated that vehicular channels will further experience the correlated scattering, i.e., non-stationarity in the frequency domain. To model the frequency non-stationarity, the authors in [14] developed a correlated tap method, where the complex correlation of amplitude and phase of taps with different delays was modeled. Also, the time non-stationarity of channels was captured by using the birth-death process method. However, in the NGSM [14], the time-frequency non-stationarity was modeled separately and individually in the time and frequency domains [15]. Furthermore, the aforementioned NGSMs in [2, 13, 14] did not explore the effect of VTDs on channel statistical properties.

Among existing modeling approaches, GBSMs have a better trade-off between the accuracy and complexity, and thus have been widely used currently in V2V channel modeling. In general, GBSMs can be further divided into regular-shaped GBSM (RS-GBSM) and irregular-shaped GBSM (IS-GBSM) [16]. RS-GBSMs utilize a regular geometry to describe the distribution of effective scatterers/clusters, and thus have low complexity. Consequently, RS-GBSMs have been extensively employed in the theoretical investigation of V2V channel modeling. For narrowband V2V channels, the proposed RS-GBSMs include a two-dimensional (2D) two-ring model [17], a 2D combination model of ellipse and two-ring [18], and a three-dimensional (3D) two-cylinder model [19]. For wideband V2V channels, a three-dimensional (3D) two concentric-cylindrical model was proposed in [20]. With the advent of fifth-generation (5G) era, a potential 5G technology, i.e., massive MIMO, is widely used in V2V communications. As indicated in the channel measurement campaign [21], the significant appearance and disappearance of clusters can be observed along the antenna array axis. Therefore, the widely used WSS assumption in the space domain is not valid in the massive MIMO scenarios, where the space non-stationarity of channels needs to be mimicked. A 3D semi-ellipsoidal V2V RS-GBSM based on the massive MIMO with uniform linear array (ULA), was proposed in [22], which characterized the time-array cluster evolution to capture the time-space non-stationarity by the birth-death process method. Considering the benefit of uniform planar array (UPA), which has a smaller size than the ULA, the UPA is more practical for massive MIMO V2V communications compared with the ULA. In [23], a 3D multi-confocal ellipsoid RS-GBSM with UPA was developed. The RS-GBSM [23] proposed a novel concept, i.e., cluster evolution area (CEA), which was the sphere with a radius of Rayleigh distance. Clusters in CEAs were modeled to perform the cluster time-space evolution based on the BD process method, and thus the channel space-time non-stationarity was modeled. Meanwhile, the impact of VTD on the channel statistical properties was investigated in [23]. However, the RS-GBSM in [23] ignored the mmWave communication, which can dramatically increase the spectrum bandwidth and has also been incorporated in V2V communications [24]. To fill this gap, the authors in [25] proposed a 3D non-stationary cylindrical V2V mmWave RS-GBSM, where the angular parameters of scatterers were resolved and further assumed to be time-varying, and thus the high delay resolution and time non-stationarity were modeled. However, the V2V RS-GBSM in [25] did not model the frequency non-stationarity of channels.

Different from RS-GBSMs, the IS-GSBM assumes that effective scatterers/clusters are placed on the irregular shapes with certain statistical distributions. In comparison with RS-GBSMs, IS-GBSMs are more accurate and realistic [6]. It is clear that, for V2V channel modeling, the IS-GBSM is a popular method, which is also extensively used in standardized channel modeling. The authors in [26] proposed a 2D V2V IS-GBSM, which mainly focused on modeling diffuse components of V2V channels. Nevertheless, the IS-GBSM in [26] is a narrowband model and ignored the channel time non-stationarity. The authors in [27] proposed a non-stationary IMT-Advanced MIMO IS-GBSM with time-variant small-scale parameters, where the birth and death behaviors of clusters were modeled by Markov chains. Paper [28] proposed a wideband V2V IS-GBSM based on measurements, and the time non-stationarity was mimicked by describing the dynamic nature of scattering environment geometry. Nevertheless, the IS-GBSMs in [27] and [28] are not applicable to mmWave/massive MIMO channels. The authors in [29] proposed a general 3D non-stationary IS-GBSM for mmWave/massive MIMO channels, where the V2V scenario was considered and the time-space non-stationarity was modeled by adopting the birth-death process to the time and space domains. However, the IS-GBSM in [29] ignored the impact of VTDs on channel statistics. To overcome this limitation, Bai et al. [30] developed a non-stationary massive MIMO V2V IS-GBSM, where a VTD-combined time-array cluster evolution algorithm was developed to model the space-time non-stationarity. This algorithm integrated the VTD into birth-death process to model the channel characteristics of V2V communications and massive MIMO jointly. The main drawback of the IS-GBSM in [30] is that the mmWave communication was ignored and the channel space-time-frequency non-stationarity was not modeled.

2.1.3 Contributions of Proposed Vehicular Channel Model

To the best of our knowledge, there is currently no channel model that can sufficiently capture the space-time-frequency non-stationarity of V2V channels, and further has the ability to explore the impact of VTDs on channel statistical properties. To fill the above gaps, we propose a novel 3D space-time-frequency non-stationary V2V IS-GBSM, which can provide some valuable guidelines for the development of standardized V2V channel models for beyond 5G (B5G) and sixth-generation (6G). The proposed IS-GBSM can be used as a simulation and validation platform, providing some guidance for the research of B5G/6G vehicular communication system algorithms. Furthermore, by modeling the channel space-time-frequency non-stationarity, the proposed IS-GBSM and the understanding of channel non-stationarity can provide some ideas and guidance for the design of the B5G/6G vehicular communication system, which is an important component in the future intelligent transportation system (ITS). The major contributions of the proposed model are outlined as follows.

1. A 3D cluster-based space-time-frequency non-stationary IS-GBSM is proposed for B5G/6G V2V mmWave massive MIMO channels. By dividing clusters into dynamic clusters and static clusters, this is the first channel model capable of investigating the impact of VTDs on channel statistics in B5G/6G V2V mmWave massive MIMO scenarios.
2. In the proposed IS-GBSM, rays within clusters are resolvable to inherently support high delay resolution. Moreover, a novel method is developed to model the space-time-frequency non-stationarity. Specifically, the dynamic correlated clusters and static correlated clusters are generated for the first time by an improved K-Means clustering algorithm, where the Elbow method is further adopted to obtain the optimum K value. Subsequently, a birth-death process method based on dynamic/static correlated clusters is adopted, which can capture the consistency in birth and death between dynamic/static correlated clusters along the time-array evolution.
3. Channel statistical properties of the proposed IS-GBSM, such as space-time-frequency correlation function (STF-CF) and Doppler power spectral density (DPSD), are derived and sufficiently studied. Simulation results demonstrate that the channel space-time-frequency non-stationarity is modeled and the VTD has

a significant impact on channel statistics. Meanwhile, close agreement between simulation results and measurements validates the utility of the proposed IS-GBSM.

2.2 A 3D Non-Stationary Vehicular Channel Model

In this section, based on the geometrical relationships, the model-related parameters of the proposed 3D non-stationary B5G/6G V2V mmWave massive MIMO IS-GBSM will be given. Furthermore, the expressions of the channel impulse response (CIR) of the proposed IS-GBSM will be derived.

2.2.1 *Model-Related Parameters*

Here, we consider a B5G/6G V2V mmWave massive MIMO communication system with L_T transmit and L_R receive antenna elements at mmWave carrier frequency f_c with a bandwidth BW. In Fig. 2.2, key parameters of the proposed IS-GBSM are presented. For clarity, some angular and distance parameters are omitted, which are similar to the given parameters. In this figure, both the Tx and Rx are mobile on the 2D plane at azimuth angles of ψ_T and ψ_R with speeds of v_T and v_R, respectively. For the Tx antenna array, θ_T and ϕ_T are used to describe the azimuth angle and elevation angle. Similarly, θ_R and ϕ_R are used to represent the azimuth angle and elevation angle for the Rx antenna array. The initial horizontal distance between Tx

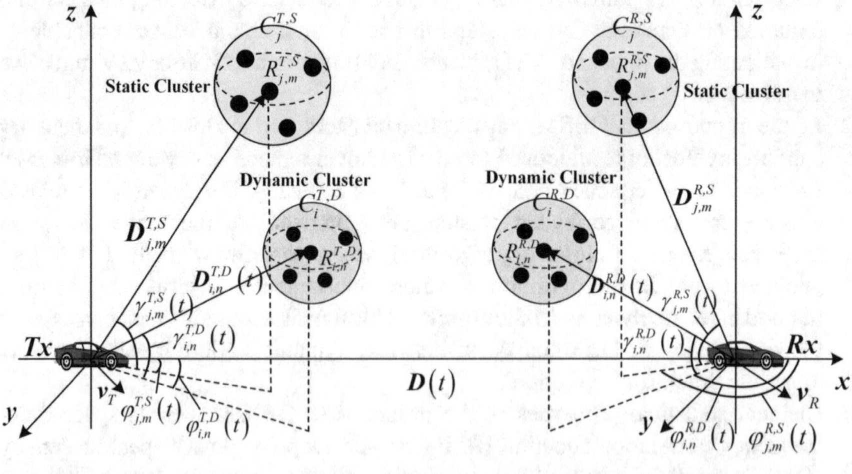

Fig. 2.2 Key parameters of the proposed 3D non-stationary IS-GBSM for B5G/6G V2V channels

and Rx is D. The adjacent antenna spacing at the Tx and Rx sides can be expressed by δ_T and δ_R. A_p^T and A_q^R denote the pth and qth antennas at the Tx and Rx sides. To explore the effect of VTDs on channel statistics, the clusters are divided into dynamic clusters and static clusters. $C_i^{T,D}$ and $C_j^{T,S}$ denote the ith dynamic and jth static clusters at the Tx side, while $C_i^{R,D}$ and $C_j^{R,S}$ denote the ith dynamic and jth static clusters at the Rx side. $D_i^{T,D}(t)$ and $D_j^{T,S}(t)$ are the distances between the Tx and the dynamic cluster $C_i^{T,D}$ and the static cluster $C_j^{T,S}$, respectively. Similarly, $D_i^{R,D}(t)$ and $D_j^{R,S}(t)$ stand for the corresponding distances at the Rx side. The azimuth angle and elevation angle of the dynamic cluster $C_i^{T,D}$ and static cluster $C_j^{T,S}$ are $\varphi_i^{T,D}(t)$ and $\gamma_j^{T,S}(t)$ at the Tx side, respectively. Similarly, $\varphi_i^{R,D}(t)$ and $\gamma_j^{R,S}(t)$ stand for the corresponding angular parameters at the Rx side. The numbers of visible dynamic clusters and visible static clusters are $I(t)$ and $J(t)$, and the movement speeds of dynamic clusters and static clusters are v_c and 0.

Figure 2.3 presents cluster types and bounce of the proposed 3D B5G/6G V2V channel model. As shown in Fig. 2.3, the proposed IS-GBSM is a twin cluster channel model, where clusters at the Tx side denote the first bounce, and clusters at the Rx side denote the last bounce. The propagation environment between clusters at the Tx side and Rx side is abstracted by a virtual link for simplicity [31]. As previously mentioned, the clusters are divided into dynamic clusters and static clusters. Their correlated clusters can be obtained by a novel method, which will be described in the sequel.

To support high delay resolution, the n-th ray within the i-th dynamic cluster and m-th ray within the j-th static cluster at the Tx/Rx side are assumed to be resolvable,

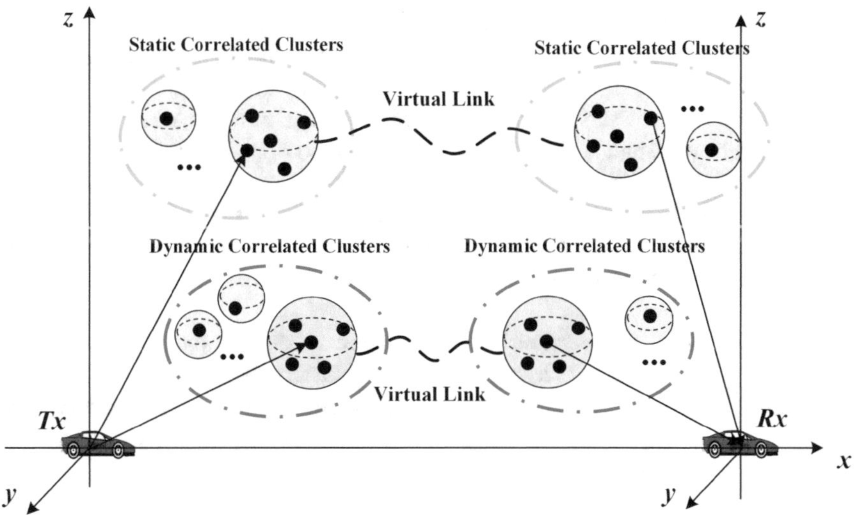

Fig. 2.3 Cluster types and bounce of the proposed 3D non-stationary B5G/6G V2V IS-GBSM

and can be represented as $R_{i,n}^{T/R,D}$ and $R_{j,m}^{T/R,S}$. The number of rays within a cluster is assumed to obey the Poisson distribution. $D_{i,n}^{T,D}(t)$ and $D_{j,m}^{T,S}(t)$ denote the distances between Tx and ray $R_{i,n}^{T,D}$ and ray $R_{j,m}^{T,S}$, respectively. Similarly, $D_{i,n}^{R,D}(t)$ and $D_{j,m}^{R,S}(t)$ denote the corresponding distances at the Rx side. The azimuth angle and elevation angle of ray $R_{i,n}^{T,D}(t)$ and ray $R_{j,m}^{T,S}$ at the Tx side are represented by $\varphi_{i,n}^{T,D}(t)$ and $\gamma_{j,m}^{T,S}(t)$. Similarly, $\varphi_{i,n}^{R,D}(t)$ and $\gamma_{j,m}^{R,S}(t)$ denote the corresponding angular parameters at the Rx side.

As mentioned before, we distinguish clusters into dynamic clusters and static clusters to investigate the impact of VTDs on channel statistics. There are some points related to dynamic and static clusters that need to be noticed. First, in V2V communication scenarios, the dynamic clusters mainly represent the moving vehicles which are at the same elevation as the Tx-Rx vehicles, while the static clusters mainly represent tall buildings on the roadside and static vehicles. The static vehicle means that the vehicle is not moving, e.g., it is in a traffic jam or waiting at a traffic light. Static vehicles can be properly represented by setting the velocity of vehicles to 0. Second, the proposed model is based on IS-GBSM and there is a geometry relationship inherent in it. Dynamic/static clusters in the environment are placed at random locations with certain statistical distributions, i.e., on irregular shapes. Based on the cluster position, other geometry information in the proposed IS-GSBM can be further derived. This is essentially different from the deterministic modeling, which needs to perform detailed measurements on actual vehicular scenarios to describe the communication environment. Third, due to the stochastic modeling, the proposed IS-GBSM does not associate a group of vehicles or a single vehicle to a dynamic/static cluster. On the contrary, the dynamic/static cluster is the equivalent cluster, not corresponding to a certain vehicle in reality.

The time-variant ratio of the numbers of dynamic and static clusters is computed as $\epsilon(t) = I(t)/J(t)$. By flexibly adjusting the ratio $\epsilon(t)$, the proposed IS-GBSM has the capability to investigate the channel statistical properties with different VTDs. In particular, it is necessary to define whether the scenario is high VTD or low VTD at the initial time instant t_0. To this end, an initial ratio $\epsilon(t_0)$ is introduced. For high VTD scenarios, the initial ratio $\epsilon(t_0)$ is set as a value greater than 1. This indicates that, at the initial time instant t_0, the number of dynamic clusters is greater than that of static clusters. In contrast, when the initial ratio $\epsilon(t_0)$ is less than or equal to 1, V2V channels can be assumed to be in low VTD scenarios at the initial time instant t_0.

To derive CIRs, the geometrical information related to dynamic and static clusters needs to be given. As dynamic and static clusters share the same calculation method, the following distance vectors utilize dynamic clusters as an instance. The distance vectors of rays $R_{i,n}^{T,D}$ and $R_{i,n}^{R,D}$ within dynamic clusters are written as

$$\mathbf{D}_{i,n}^{T,D}(t) = D_{i,n}^{T,D}(t) \begin{bmatrix} \cos \gamma_{i,n}^{T,D}(t) \cos \varphi_{i,n}^{T,D}(t) \\ \cos \gamma_{i,n}^{T,D}(t) \sin \varphi_{i,n}^{T,D}(t) \\ \sin \gamma_{i,n}^{T,D}(t) \end{bmatrix}^{\mathrm{T}} \tag{2.1}$$

$$\mathbf{D}_{i,n}^{R,D}(t) = D_{i,n}^{R,D}(t) \begin{bmatrix} \cos \gamma_{i,n}^{R,D}(t) \cos \varphi_{i,n}^{R,D}(t) \\ \cos \gamma_{i,n}^{R,D}(t) \sin \varphi_{i,n}^{R,D}(t) \\ \sin \gamma_{i,n}^{R,D}(t) \end{bmatrix}^{\mathrm{T}} + \mathbf{D}(t). \tag{2.2}$$

Obviously, the distance vectors $\mathbf{D}_{j,m}^{T,S}(t)$ and $\mathbf{D}_{j,m}^{R,S}(t)$ of rays within static clusters are not time-variant, and thus can be simplified as $\mathbf{D}_{j,m}^{T,S}$ and $\mathbf{D}_{j,m}^{R,S}$. Moreover, the distance vectors of dynamic clusters $C_i^{T,D}$ and $C_i^{R,D}$ are written as

$$\mathbf{D}_{i}^{T,D}(t) = D_{i}^{T,D}(t) \begin{bmatrix} \cos \gamma_{i}^{T,D}(t) \cos \varphi_{i}^{T,D}(t) \\ \cos \gamma_{i}^{T,D}(t) \sin \varphi_{i}^{T,D}(t) \\ \sin \gamma_{i}^{T,D}(t) \end{bmatrix}^{\mathrm{T}} \tag{2.3}$$

$$\mathbf{D}_{i}^{R,D}(t) = D_{i}^{R,D}(t) \begin{bmatrix} \cos \gamma_{i}^{R,D}(t) \cos \varphi_{i}^{R,D}(t) \\ \cos \gamma_{i}^{R,D}(t) \sin \varphi_{i}^{R,D}(t) \\ \sin \gamma_{i}^{R,D}(t) \end{bmatrix}^{\mathrm{T}} + \mathbf{D}(t). \tag{2.4}$$

Similarly, the distance vectors of static clusters $C_j^{T,S}$ and $C_j^{R,S}$ are $\mathbf{D}_j^{T,S}$ and $\mathbf{D}_j^{R,S}$. Moreover, the distance vectors of pth Tx antenna and qth Rx antenna are given by

$$\mathbf{D}_p^T = \frac{L_T - 2p + 1}{2} \delta_T \begin{bmatrix} \cos \phi_T \cos \theta_T \\ \cos \phi_T \sin \theta_T \\ \sin \phi_T \end{bmatrix}^{\mathrm{T}} \tag{2.5}$$

$$\mathbf{D}_q^R(t) = \frac{L_R - 2q + 1}{2} \delta_R \begin{bmatrix} \cos \phi_R \cos \theta_R \\ \cos \phi_R \sin \theta_R \\ \sin \phi_R \end{bmatrix}^{\mathrm{T}} + \mathbf{D}(t). \tag{2.6}$$

The distance vectors between the rays $R_{i,n}^{T,D}$ and $R_{i,n}^{R,D}$ within dynamic clusters and antenna elements are given as

$$\mathbf{D}_{p,i,n}^{T,D}(t) = \mathbf{D}_{i,n}^{T,D}(t) - \mathbf{D}_p^T \tag{2.7}$$

$$\mathbf{D}_{q,i,n}^{R,D}(t) = \mathbf{D}_{i,n}^{R,D}(t) - \mathbf{D}_q^R(t). \tag{2.8}$$

The distance vectors $\mathbf{D}_{p,j,m}^{T,S}(t)$ and $\mathbf{D}_{q,j,m}^{R,S}(t)$ between rays within static clusters and antenna elements can be similarly obtained. The distance vectors between the dynamic clusters $C_i^{T,D}$ and $C_i^{R,D}$ and antenna elements are expressed as

$$\mathbf{D}_{p,i}^{T,D}(t) = \mathbf{D}_i^{T,D}(t) - \mathbf{D}_p^T \tag{2.9}$$

$$\mathbf{D}_{q,i}^{R,D}(t) = \mathbf{D}_i^{R,D}(t) - \mathbf{D}_q^R(t). \tag{2.10}$$

Also, the distance vectors $\mathbf{D}_{p,j}^{T,S}(t)$ and $\mathbf{D}_{q,j}^{R,S}(t)$ can be obtained analogously to (9) and (10). Furthermore, the LoS distance vector is given by

$$\mathbf{D}_{qp}^{LoS}(t) = \mathbf{D}_q^R(t) - \mathbf{D}_p^T. \tag{2.11}$$

Based on the distance vectors related to the clusters, the delays of LoS component $\tau^{LoS}(t)$, dynamic clusters $\tau_i^D(t)$, and static clusters $\tau_j^S(t)$ can be given as

$$\tau^{LoS}(t) = \frac{\|\mathbf{D}(t)\|}{c} \tag{2.12}$$

$$\tau_i^D(t) = \frac{\left\|\mathbf{D}_i^{T,D}(t)\right\| + \left\|\mathbf{D}_i^{R,D}(t)\right\|}{c} + \tilde{\tau}_i^D(t) \tag{2.13}$$

$$\tau_j^S(t) = \frac{\left\|\mathbf{D}_j^{T,S}(t)\right\| + \left\|\mathbf{D}_j^{R,S}(t)\right\|}{c} + \tilde{\tau}_j^S(t) \tag{2.14}$$

$$\begin{aligned} h_{qp}(t,\tau) = & \underbrace{\sqrt{\frac{\Omega}{\Omega+1}} h_{qp}^{LoS}(t)\delta\left(\tau - \tau^{LoS}(t)\right)}_{\text{LoS}} \\ & + \underbrace{\sqrt{\frac{\eta_1}{\Omega+1}} \sum_{i=1}^{I(t)} \sum_{n=1}^{N(t)} h_{qp,i,n}^{NLoS,D}(t)\delta\left(\tau - \tau_i^D(t) - \tau_{i,n}^D(t)\right)}_{\text{NLoS, Dynamic Clusters}} \\ & + \underbrace{\sqrt{\frac{\eta_2}{\Omega+1}} \sum_{j=1}^{J(t)} \sum_{m=1}^{M(t)} h_{qp,j,m}^{NLoS,S}(t)\delta\left(\tau - \tau_j^S(t) - \tau_{j,m}^S(t)\right)}_{\text{NLoS, Static Clusters}} \end{aligned}$$

where

$$\begin{aligned} h_{qp}^{LoS}(t) = & \begin{bmatrix} F_{p,V}^T(\mathbf{D}_{qp}^{LoS}(t), \mathbf{D}_p^T(t)) \\ F_{p,H}^T(\mathbf{D}_{qp}^{LoS}(t), \mathbf{D}_p^T(t)) \end{bmatrix}^{\mathrm{T}} \begin{bmatrix} e^{j\Psi_{LoS}} & 0 \\ 0 & -e^{j\Psi_{LoS}} \end{bmatrix} \begin{bmatrix} F_{q,V}^R(\mathbf{D}_{qp}^{LoS}(t), \mathbf{D}_q^R(t)) \\ F_{q,H}^R(\mathbf{D}_{qp}^{LoS}(t), \mathbf{D}_q^R(t)) \end{bmatrix} \\ & \times e^{j2\pi f_{qp}^{LoS}(t)t} e^{j\Phi_0} e^{j\frac{2\pi}{\lambda}\left\|\mathbf{D}_{qp}^{LoS}(t)\right\|} \end{aligned} \tag{2.15}$$

$$\begin{aligned} h_{qp,i,n}^{NLoS,D}(t) = & \begin{bmatrix} F_{p,V}^T(\mathbf{D}_{p,i,n}^{T,D}(t), \mathbf{D}_p^T(t)) \\ F_{p,H}^T(\mathbf{D}_{p,i,n}^{T,D}(t), \mathbf{D}_p^T(t)) \end{bmatrix}^{\mathrm{T}} \begin{bmatrix} e^{j\Psi_{D,i,n}^{VV}} & \sqrt{\kappa} e^{j\Psi_{D,i,n}^{VH}} \\ \sqrt{\kappa} e^{j\Psi_{D,i,n}^{HV}} & e^{j\Psi_{D,i,n}^{HH}} \end{bmatrix} \begin{bmatrix} F_{q,V}^R(\mathbf{D}_{q,i,n}^{R,D}(t), \mathbf{D}_q^R(t)) \\ F_{q,H}^R(\mathbf{D}_{q,i,n}^{R,D}(t), \mathbf{D}_q^R(t)) \end{bmatrix} \\ & \times \sqrt{\bar{P}_{i,n}^D(t)} e^{j2\pi f_{pi,n}^{T,D}(t)t} e^{j2\pi f_{qi,n}^{R,D}(t)t} e^{j\Phi_0} e^{j\frac{2\pi}{\lambda}\left[\left\|\mathbf{D}_{p,i,n}^{T,D}(t)\right\| + \left\|\mathbf{D}_{q,i,n}^{R,D}(t)\right\|\right]} \end{aligned} \tag{2.16}$$

$$h_{qp,j,m}^{NLoS,S}(t) = \begin{bmatrix} F_{p,V}^{T}(\mathbf{D}_{p,j,m}^{T,S}(t), \mathbf{D}_{p}^{T}(t)) \\ F_{p,H}^{T}(\mathbf{D}_{p,j,m}^{T,S}(t), \mathbf{D}_{p}^{T}(t)) \end{bmatrix}^{\mathrm{T}} \begin{bmatrix} e^{j\Psi_{S,j,m}^{VV}} & \sqrt{\kappa}e^{j\Psi_{S,j,m}^{VH}} \\ \sqrt{\kappa}e^{j\Psi_{S,j,m}^{HV}} & e^{j\Psi_{S,j,m}^{HH}} \end{bmatrix} \begin{bmatrix} F_{q,V}^{R}(\mathbf{D}_{q,j,m}^{R,S}(t), \mathbf{D}_{q}^{R}(t)) \\ F_{q,H}^{R}(\mathbf{D}_{q,j,m}^{R,S}(t), \mathbf{D}_{q}^{R}(t)) \end{bmatrix}$$
$$\times \sqrt{\bar{P}_{j,m}^{S}(t)} e^{j2\pi f_{pj,m}^{T,S}(t)t} e^{j2\pi f_{qj,m}^{R,S}(t)t} e^{j\Phi_0} e^{j\frac{2\pi}{\lambda}\left[\left\|\mathbf{D}_{p,j,m}^{T,S}(t)\right\| + \left\|\mathbf{D}_{q,j,m}^{R,S}(t)\right\|\right]}. \tag{2.17}$$

---

where $c = 3 \times 10^8$ m/s is the speed of light, and $\tilde{\tau}_i^D(t)$ and $\tilde{\tau}_j^S(t)$ further denote the virtual delays caused by the virtual link of dynamic clusters and static clusters, respectively. Typically, the virtual delays $\tilde{\tau}_i^D(t)$ and $\tilde{\tau}_j^S(t)$ can be assumed to obey the exponential distribution.

2.2.2 Channel Impulse Response

The proposed B5G/6G V2V mmWave massive MIMO IS-GBSM at time t with delay τ can be described by a matrix $\mathbf{H}(t, \tau) = [h_{qp}(t, \tau)]_{L_R \times L_T}$ with $q = 1, 2, ..., L_R$ and $p = 1, 2, ..., L_T$. The entries of $\mathbf{H}(t, \tau)$ are a superposition of the LoS component and the NLoS component. Note that, since we distinguish clusters into dynamic clusters and static clusters, the NLoS component needs to be further divided into links from dynamic clusters and static clusters. Therefore, the CIR $h_{qp}(t, \tau)$ for the Tx and Rx antenna element pair p and q can be written as (2.15)–(2.17), shown at the top of next page. In (2.15), Ω denotes the Ricean factor, which is assumed to be constant during the generation of channel coefficients for simplicity. η_1 and η_2 are the parameters to represent the power ratio of dynamic clusters and static clusters, satisfying $\eta_1 + \eta_2 = 1$. $N(t)$ and $M(t)$ are the numbers of rays within dynamic clusters and static clusters, respectively. $\tau_{i,n}^D(t)$ and $\tau_{j,m}^S(t)$ denote the relative delays of ray $R_{i,n}^D$ and ray $R_{j,m}^S$, respectively. In (2.15)–(2.17), functions $F_V^R(\cdot)$, $F_H^R(\cdot)$, $F_V^T(\cdot)$, and $F_H^T(\cdot)$ are the antenna patterns in the global ordinate system (GCS) with the origin at the center of Rx array. Furthermore, Ψ_{LoS} follows the uniform distribution in $(0, 2\pi]$, and $\Psi_{D,i,n}^{VV}$, $\Psi_{D,i,n}^{VH}$, $\Psi_{D,i,n}^{HV}$, $\Psi_{D,i,n}^{HH}$ denote the initial random phases of the nth ray within the ith dynamic cluster in four polarization directions. Similarly, $\Psi_{S,j,m}^{VV}$, $\Psi_{S,j,m}^{VH}$, $\Psi_{S,j,m}^{HV}$, $\Psi_{S,j,m}^{HH}$ are the initial random phases of the mth ray within the jth static cluster. Also, the above eight phases are uniformly distributed in $(0, 2\pi]$. Φ_0 denotes the initial phase. Moreover, κ represents the cross polarization power ration, and $\bar{P}_{i,n}^D(t)$ and $\bar{P}_{j,m}^S(t)$ are the normalized mean power of rays within dynamic clusters and static clusters. The Doppler frequency of the LoS component is $f_{qp}^{LoS}(t) = \frac{1}{\lambda} \frac{\langle \mathbf{D}_{qp}^{LoS}(t), \mathbf{v}_R - \mathbf{v}_T \rangle}{\left\|\mathbf{D}_{qp}^{LoS}(t)\right\|}$. In addition, Doppler frequencies of the NLoS component can be written as $f_{p,i,n}^{T,D}(t) = \frac{\langle \mathbf{D}_{p,i,n}^{T,D}(t), \mathbf{v}_T - \mathbf{v}_c \rangle}{\lambda \left\|\mathbf{D}_{p,i,n}^{T,D}(t)\right\|}$, $f_{q,i,n}^{R,D}(t) = \frac{\langle \mathbf{D}_{q,i,n}^{R,D}(t), \mathbf{v}_R - \mathbf{v}_c \rangle}{\lambda \left\|\mathbf{D}_{q,i,n}^{R,D}(t)\right\|}$, $f_{p,j,m}^{T,S}(t) = \frac{\langle \mathbf{D}_{p,j,m}^{T,S}(t), \mathbf{v}_T \rangle}{\lambda \left\|\mathbf{D}_{p,j,m}^{T,S}(t)\right\|}$, and $f_{q,j,m}^{R,S}(t) = \frac{\langle \mathbf{D}_{q,j,m}^{R,S}(t), \mathbf{v}_R \rangle}{\lambda \left\|\mathbf{D}_{q,j,m}^{R,S}(t)\right\|}$, where $\langle \cdot, \cdot \rangle$ is the inner product operator.

Table 2.1 lists the key parameters of the proposed B5G/6G V2V mmWave massive MIMO IS-GBSM. In this section, we have derived key model-related parameters and have given the CIRs of the proposed IS-GBSM. In the following paragraphs, we will develop a novel method to sufficiently model the space-time-frequency non-stationarity of B5G/6G V2V mmWave massive MIMO channels.

Table 2.1 Definitions of Key B5G/6G V2V IS-GBSM parameters

Symbol	Definition
δ_T, δ_R	Adjacent antenna spacing at the Tx and Rx sides
K^D, K^S	Optimum K values for dynamic and static correlated clusters
$I(t), J(t)$	Numbers of dynamic and static clusters
$N(t), M(t)$	Numbers of rays in dynamic and static clusters
$\bar{P}^D_{i,n}(t), \bar{P}^S_{j,m}(t)$	Normalized mean power of $R^D_{i,n}$ and $R^S_{j,m}$
$\lambda_B^{D/S}, \lambda_D^{D/S}$	Birth (generation) rate and death (recombination) rate of dynamic/static clusters
$\boldsymbol{\rho}^D(t,f), \boldsymbol{\rho}^S(t,f)$	Correlation coefficient matrices of dynamic and static correlated clusters
$\mathbf{D}^{LoS}_{qp}(t)$	Distance vector between the pth Tx antenna and the qth Rx antenna
$\mathbf{D}^T_p, \mathbf{D}^R_q$	Distance vectors of the pth Tx antenna and the qth Rx antenna
$\mathbf{D}^{T/R,D}_{p/q,i}(t), \mathbf{D}^{T/R,S}_{p/q,j}(t)$	Distance vectors between $C_i^{T/R,D}$ and $C_j^{T/R,S}$ and p/qth Tx/Rx antennas
$\mathbf{D}^{T/R,D}_{p/q,i,n}(t), \mathbf{D}^{T/R,S}_{p/q,j,m}(t)$	Distance vectors between $R^{T/R,D}_{i,n}$ and $R^{T/R,S}_{j,m}$ and p/qth Tx/Rx antennas
$\mathbf{v}_T, \mathbf{v}_R$	Velocity vectors of Tx array and Rx array
$\mathbf{v}_c, 0$	Velocity vectors of dynamic and static clusters
$\epsilon(t)$	Ratio of the number of dynamic clusters and static clusters
$\tau^{LoS}(t)$	LoS delay of the Tx and Rx
$\tau_i^D(t), \tau_j^S(t)$	Delays of C_i^D and C_j^S
$\tilde{\tau}_i^D(t), \tilde{\tau}_j^S(t)$	Virtual delays of C_i^D and C_j^S
$\tau^D_{i,n}(t), \tau^S_{j,m}(t)$	Relative delays of $R^D_{i,n}$ and $R^S_{j,m}$
$f^{LoS}_{qp}(t)$	LoS Doppler frequency between the pth Tx antenna and the qth Rx antenna
$f^{T,D}_{pi,n}(t), f^{R,D}_{qi,n}(t)$	Doppler frequencies of pth Tx antenna and qth Rx antenna via $R^{T/R,D}_{i,n}$
$f^{T,S}_{pj,m}(t), f^{R,S}_{qj,m}(t)$	Doppler frequencies of pth Tx antenna and qth Rx antenna via $R^{T/R,S}_{j,m}$
$\theta_{T/R}, \phi_{T/R}$	Azimuth and elevation angles of the Tx/Rx array broadside
$\varphi_i^{T/R,D}(t), \gamma_j^{T/R,S}(t)$	Azimuth and elevation angles between $C_i^{T/R,D}$ and $C_j^{T/R,S}$ and Tx/Rx center
$\varphi^{T/R,D}_{i,n}(t), \gamma^{T/R,S}_{j,m}(t)$	Azimuth and elevation angles between $R^{T/R,D}_{i,n}$ and $R^{T/R,S}_{j,m}$ and Tx/Rx center

2.3 Vehicular Channel Space-Time-Frequency Non-stationary Modeling

In the proposed IS-GBSM, a novel method will be developed to capture the channel space-time-frequency non-stationarity. In the developed method, an improved K-Means clustering algorithm is employed to acquire the dynamic correlated clusters and the static correlated clusters, where the Elbow method is further used to find the optimum K value. Then, in the time-array evolution, a novel birth-death process based on dynamic/static correlated clusters is adopted to model the consistency in birth and death between dynamic/static correlated clusters. Therefore, the channel space-time-frequency non-stationarity is modeled sufficiently.

2.3.1 Generation of Dynamic Correlated Clusters and Static Correlated Clusters

In tap delay line (TDL) channel models, the frequency non-stationarity can be modeled by obtaining correlated taps with different delays [2]. This is reasonable because frequency non-stationary channels are equivalent to correlated scattering channels, where attenuation and phase shift with paths in the different delays are correlated [32]. As an evolution of the TDL channel model, it is natural that cluster-based channel models can generate correlated clusters with different delays to model the frequency non-stationarity. Nevertheless, different from taps, which are solely restricted to differences in the delay, clusters differ in the delay and location. In this case, both the difference in delays and locations have an impact on the generation of correlated clusters. Furthermore, it can be expected that the more different the delay and location of clusters, the more likely they are uncorrelated clusters.

To obtain the correlated clusters that have similar locations, an improved K-Means clustering algorithm is employed, where the Elbow method is further adopted to obtain the optimum K value. The K-Means clustering algorithm is widely employed to categorize sample points into K sets. Note that the sample points in the same set are close to each other [33]. By applying this algorithm to clusters in the propagation environment, they will be classified into K sets according to their locations. Clusters within the same set have closely spaced locations. Here, we define that clusters within the same set are correlated to each other, i.e., correlated clusters.

As previously mentioned, clusters are divided into dynamic clusters and static clusters. In this case, it needs to generate the dynamic correlated clusters together with static correlated clusters. To this end, the K-Means clustering algorithm is employed in dynamic clusters and static clusters respectively. For K-Means clustering algorithm, it is well known that the performance of this algorithm significantly depends on the K value. Furthermore, it is also difficult to choose a proper K value. Therefore, obtaining an optimum K value for the K-Means clustering algorithm is essential and challenging. To optimally classify the dynamic clusters and static clusters into K^D sets and K^S sets, an improved K-Means clustering algorithm is employed. In this improved algorithm, the Elbow method is adopted to acquire the optimum K value.

Based on [34], steps of the improved K-Means clustering algorithm can be summarized as followed. Since dynamic and static clusters have a similar procedure, we also take the dynamic cluster as an example.

Step 1: Randomly select K^D dynamic clusters in the communication environment as the initial centroid.

Step 2: For other dynamic clusters, find the nearest centroid according to the Euclidean distance, and properly classify them into the specific set where the nearest centroid is located.

Step 3: These dynamic clusters in the communication environment are successfully divided into K^D sets, and then recalculate the centroid of each set.

Step 4: Repeat steps 2 and 3 until the centroids are no longer changed. Then, the clusters have been classified into K^D sets, and the dynamic correlated clusters can be obtained. At the same time, calculate the sum of square error (SSE) between each dynamic cluster and its corresponding centroid of the set.

$$SSE^D = \sum_{k^d=1}^{K^D} \sum_{i=1}^{I(t)} \left\| P_i^D - \mu_{k^d}^D \right\|_2^2 \tag{2.18}$$

where P_i^D is the 3D position vector of the ith dynamic cluster, and $\mu_{k^d}^D$ is the 3D position vector of the centroid of the k^dth set of dynamic correlated clusters. The operator $\|\cdot\|_2^2$ represents the square of the Euclidean norm.

Step 5: After obtaining dynamic correlated clusters, we use the Elbow method to find the optimum K^D. Starting from $K^D = 1$ and keep increasing it in each step by 1, i.e., $K'^D = K^D + 1$. According to (2.18), compute the SSE for each K^D. Based on each K^D and calculated SSE, the slope $k_{K^D-1}^{SSE}$ can be obtained according to $K^D - 1$ and its SSE. Similarly, the slope $k_{K^D+1}^{SSE}$ can be obtained according to $K^D + 1$ and its SSE. Compute the absolute slope difference between the two, i.e., $\Delta k_{K^D}^{SSE} = \left| k_{K^D+1}^{SSE} - k_{K^D-1}^{SSE} \right|$. Note that the K^D at which the maximum value of the slope difference $\Delta k_{K^D}^{SSE}$ is located is the optimum value of K^D, i.e., elbow-point, where the value of SSE drops drastically and forms a small angle [35].

Static correlated clusters with the optimum value of K^S can be similarly generated. To show the improved K-Means clustering algorithm, we take the high VTD scenario with initially 20 dynamic clusters and 5 static clusters as an example. From Fig. 2.4, the optimum K values for dynamic clusters and static clusters are $K^D = 3$ and $K^S = 2$, respectively. It is the first time that the improved K-Means

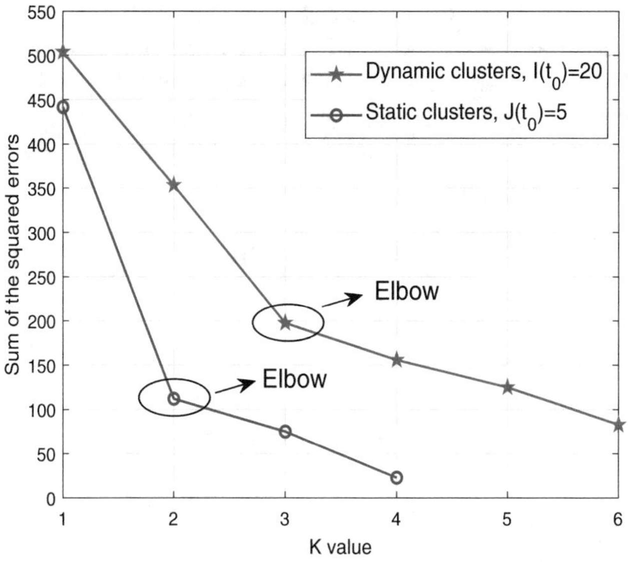

Fig. 2.4 Optimum K values for dynamic correlated clusters and static correlated clusters

clustering algorithm is applied to V2V channel modeling to acquire the dynamic and static correlated clusters with optimum K values.

After obtaining dynamic and static correlated clusters, the magnitude of correlation of clusters needs to be described quantitatively. Similar to correlated taps, the correlation coefficient is inversely proportional to the relative delays of clusters. Furthermore, [36] indicated that a ray with larger frequency exhibits larger dependency, and hence the influence of frequency on the correlation coefficient should be considered. As a result, the correlation coefficient of dynamic clusters $\rho^D_{ii'}(t, f)$ and static clusters $\rho^S_{jj'}(t, f)$ can be given by

$$\rho^D_{ii'}(t, f) = \frac{\beta^D}{\left|\tau^D_i(t) - \tau^D_{i'}(t)\right|}(\frac{f}{f_c})^{\eta^D}, \forall i, i' \in I(t), i \neq i' \tag{2.19}$$

$$\rho^S_{jj'}(t, f) = \frac{\beta^S}{\left|\tau^S_j(t) - \tau^S_{j'}(t)\right|}(\frac{f}{f_c})^{\eta^S}, \forall j, j' \in J(t), j \neq j' \tag{2.20}$$

where $\beta^{D/S}$ is a correlation factor for dynamic/static clusters, making the largest correlation coefficient less than 1. In this case, the correlation coefficient is ranging from 0 to 1. $\tau^D_{i/i'}(t)$ and $\tau^S_{j/j'}(t)$ are the delays of i/i'th dynamic clusters and j/j'th static clusters, respectively. $\eta^{D/S}$ is the frequency-dependent factor to measure the degree of frequency correlation of dynamic/static clusters [37]. Note that there is no correlation between dynamic clusters and static clusters, and thus the correlation coefficient between dynamic clusters and static clusters is identically vanishing. This is reasonable because locations of dynamic clusters and static clusters are quite different. For $i = i'$ and $j = j'$, the correlation coefficient $\rho_{ii'}^{D}(t, f)$ and $\rho_{jj'}^{S}(t, f)$ are auto-correlation coefficient, thus identically equal to 1, i.e., $\rho_{ii/i'i'}^{D}(t, f) \equiv 1$ and $\rho_{jj/j'j'}^{S}(t, f) \equiv 1$. Accordingly, the correlation coefficient matrix of dynamic/static clusters $\boldsymbol{\rho}^{D/S}(t, f)$ is written as

$$\boldsymbol{\rho}^{D/S}(t, f) = \begin{bmatrix} 1 & \rho_{1,2}^{D/S}(t, f) & \rho_{1,3}^{D/S}(t, f) & \cdots & \rho_{1,i/j}^{D/S}(t, f) \\ \rho_{2,1}^{D/S}(t, f) & 1 & \rho_{2,3}^{D/S}(t, f) & \cdots & \rho_{2,i/j}^{D/S}(t, f) \\ \rho_{3,1}^{D/S}(t, f) & \rho_{3,2}^{D/S}(t, f) & 1 & \cdots & \rho_{3,i/j}^{D/S}(t, f) \\ \vdots & \vdots & \vdots & \ddots & \vdots \\ \rho_{i/j,1}^{D/S}(t, f) & \rho_{i/j,2}^{D/S}(t, f) & \rho_{i/j,3}^{D/S}(t, f) & \cdots & 1 \end{bmatrix}. \tag{2.21}$$

Note that the aforementioned correlation coefficient matrix $\boldsymbol{\rho}^{D/S}(t, f)$ is a real symmetric matrix. In addition, since the diagonal elements denote the auto-correlation coefficient, they are equal to 1.

2.3.2 *Time-Array Evolution of Dynamic Correlated Clusters and Static Correlated Clusters*

The impact of generating dynamic and static correlated clusters on time-array evolution will be discussed sufficiently in this part. Here, it is assumed that $\lambda_B^{D/S}$ and $\lambda_D^{D/S}$ are the birth (generation) rate and death (recombination) rate of dynamic/static clusters, respectively.

2.3.2.1 Initialization of Correlated Cluster Sets

The first step is the initialization of correlated cluster sets at the initial time instant t_0. For dynamic clusters, the initial number is $I(t_0)$. Therefore, $I(t_0)$ dynamic clusters are randomly generated at the Tx side and Rx side, respectively. These initial dynamic clusters are assumed to be visible at the Tx and Rx sides, and can be represented as $\left\{C_i^{T,D} : i = 1, 2, ..., I(t_0)\right\}$ and $\left\{C_i^{R,D} : i = 1, 2, ..., I(t_0)\right\}$. Then, the key parameters of rays and the dynamic clusters need to be determined by statistical distributions. Specifically, the number of rays within a dynamic cluster follows the Poisson distribution [38], the delay and mean power of dynamic clusters follow the exponential distribution [39], and angular parameters of dynamic clusters follow the Wrapped Gaussian [29]. Similarly, $J(t_0)$ visible static clusters at the Tx and Rx sides are generated, and statistical distributions of key parameters related to static clusters are the same as those of dynamic clusters.

2.3.2.2 Array Evolution of Correlated Clusters

The second step is the array evolution of correlated clusters at the initial time instant t_0, and the space non-stationarity can be modeled. Also, the birth-death process is used to model the array evolution. For dynamic clusters, the survival probabilities along the array evolution at the Tx side $P^{T,D}_{survival}$ and Rx side $P^{R,D}_{survival}$ are modeled as exponential functions [40], and are written by

$$P^{T,D}_{survival} = e^{-\lambda^D_D \frac{\delta_T}{D^a_{c,D}}} \tag{2.22}$$

$$P^{R,D}_{survival} = e^{-\lambda^D_D \frac{\delta_R}{D^a_{c,D}}} \tag{2.23}$$

where $D^a_{c,D}$ is defined as the antenna spacing correlation coefficient for dynamic clusters. Note that the survival probabilities of static clusters $P^{T,S}_{survival}$ and $P^{R,S}_{survival}$ can be derived similarly to dynamic clusters.

In addition to the birth-death process, the cluster visibility region (VR) also has the ability to model the channel space non-stationarity. In this method, the clusters that are located nearby tend to have the same visibility status, which is consistent with reality. As mentioned before, the correlated clusters have similar locations. In this case, when a cluster is visible to an antenna element, its correlated clusters may also be visible to the antenna element, and vice versa. Furthermore, it can be expected that, if the correlation coefficient between two correlated clusters is large, the probability that they have the same visibility status is also large. In other words, the degree to which the visibility status of a cluster is influenced by its correlated clusters is proportional to their correlation coefficient. Here, we termed this phenomenon as *consistency in birth and death between correlated clusters*. To model this phenomenon, a novel birth-death process based on dynamic/static correlated clusters is developed and used in the cluster array evolution. As a result, the effect of introducing correlated clusters on survival probabilities is fully considered. Also for the dynamic clusters, based on the correlation coefficient, the survival probabilities of the ith dynamic clusters $P^{T,D}_{i,survival}(t_0)$ and $P^{R,D}_{i,survival}(t_0)$ are expressed as

$$\sum_{i'=1}^{I(t_0)} \left[x^{T,D}_{i'} \rho^D_{ii'}(t_0, f) e^{-\lambda^D_D \frac{\delta_T}{D^a_{c,D}}} + y^{T,D}_{i'} \rho^D_{ii'}(t_0, f) \left(1 - e^{-\lambda^D_D \frac{\delta_T}{D^a_{c,D}}} \right) \right] \tag{2.24}$$

$$\sum_{i'=1}^{I(t_0)} \left[x^{R,D}_{i'} \rho^D_{ii'}(t_0, f) e^{-\lambda^D_D \frac{\delta_R}{D^a_{c,D}}} + y^{R,D}_{i'} \rho^D_{ii'}(t_0, f) \left(1 - e^{-\lambda^D_D \frac{\delta_R}{D^a_{c,D}}} \right) \right] \tag{2.25}$$

where $x^{T/R,D}_{i'}$ and $y^{T/R,D}_{i'}$ are the status-dependent factors of the i'th dynamic correlated clusters at the Tx/Rx side for array evolution. In the array evolution, when the i'th dynamic cluster is in the death status at the Tx/Rx side, we have $x_{i'}^{T/R,D} = 1$ and $y_{i'}^{T/R,D} = 0$. In contrast, when the i'th dynamic cluster is in the birth status at the Tx/Rx side, we have $x_{i'}^{T/R,D} = 0$ and $y_{i'}^{T/R,D} = -1$. Accordingly, the status-dependent factors $x_{i'}^{T/R,D}$ and $y_{i'}^{T/R,D}$ satisfy

$$\begin{cases} x_{i'}^{T/R,D} = 1, \ y_{i'}^{T/R,D} = 0 \ cluster \to birth \\ x_{i'}^{T/R,D} = 0, \ y_{i'}^{T/R,D} = -1 \ cluster \to death \end{cases}$$

Therefore, if the correlation coefficient between the ith dynamic cluster and the i'th dynamic cluster is not 0, when the i'th dynamic cluster is in the birth status, the ith dynamic cluster has a higher survival probability. This makes the ith dynamic cluster more likely to be a visible cluster. Conversely, the disappearance of i'th dynamic correlated cluster will lead to a decrease in the survival probability of ith dynamic cluster. Consequently, (25) and (26) can properly describe the consistency in birth and death between correlated clusters in the space domain. Similarly, the survival probabilities $P_{j,survival}^{T,S}(t_0)$ and $P_{j,survival}^{R,S}(t_0)$ of the jth static cluster at the Tx side and Rx side can be given by

$$P_{j,survival}^{T,S}(t_0) = \sum_{j'=1}^{J(t_0)} \left[x_{j'}^{T,S} \rho_{jj'}^{S}(t_0, f) e^{-\lambda_D^S \frac{\delta_T}{D_{c,S}^a}} + y_{i'}^{T,S} \rho_{jj'}^{S}(t_0, f) \left(1 - e^{-\lambda_D^S \frac{\delta_T}{D_{c,S}^a}}\right) \right] \tag{2.26}$$

$$P_{j,survival}^{R,S}(t_0) = \sum_{j'=1}^{J(t_0)} \left[x_{j'}^{R,S} \rho_{jj'}^{S}(t_0, f) e^{-\lambda_D^S \frac{\delta_R}{D_{c,S}^a}} + y_{i'}^{R,S} \rho_{jj'}^{S}(t_0, f) \left(1 - e^{-\lambda_D^S \frac{\delta_R}{D_{c,S}^a}}\right) \right] \tag{2.27}$$

where $x_{j'}^{T/R,S}$ and $y_{j'}^{T/R,S}$ are the status-dependent factors of static correlated clusters at the Tx/Rx side for cluster array evolution, and have the same value rules as $x_{i'}^{T/R,D}$ and $y_{i'}^{T/R,D}$.

By modeling the consistency in birth and death between correlated clusters along the array axis, clusters that are located nearby tend to have the same status. In addition, the mean numbers of newly generated dynamic clusters and static clusters at the Tx/Rs side are expressed as

$$E\left[I_{new}^{T/R,D}\right] = \frac{\lambda_B^D}{\lambda_D^D}\left(1 - \sum_{i=1}^{I(t_0)} P_{i,survival}^{T/R,D}(t_0)/I(t_0)\right) \tag{2.28}$$

$$E\left[J_{new}^{T/R,S}\right] = \frac{\lambda_B^S}{\lambda_D^S}\left(1 - \sum_{j=1}^{J(t_0)} P_{j,survival}^{T/R,S}(t_0)/J(t_0)\right) \tag{2.29}$$

where E[$\cdot$] designates the expectation operator.

2.3.2.3 Time Evolution of Correlated Clusters

The third step is the time evolution on the time axis, and the time non-stationarity can be modeled. For dynamic clusters, the survival probability $P^{D,time}_{survival}$ after a time step Δt can be given by

$$P^{D,time}_{survival} = e^{-\lambda^{D}_{D}\frac{(\Delta v^{D}_{T}+\Delta v^{D}_{R})\Delta t}{D^{S}_{c,D}}} \tag{2.30}$$

where Δv^{D}_{T} and Δv^{D}_{R} are the relative velocities, which are computed as $\Delta v^{D}_{T} = \|\mathbf{v}_{T} - \mathbf{v}_{c}\|$ and $\Delta v^{D}_{R} = \|\mathbf{v}_{R} - \mathbf{v}_{c}\|$. $D^{s}_{c,D}$ is a scenario-dependent factor to represent the degree of spatial correlation for dynamic clusters. The survival probability of static clusters $P^{S,time}_{survival}$ can be derived by substituting D by S in (2.30). Since the velocity of static clusters is zero, the relative velocities between static clusters and Tx/Rx are computed as $\Delta v^{S}_{T} = \|\mathbf{v}_{T}\|$ and $\Delta v^{S}_{R} = \|\mathbf{v}_{R}\|$, respectively.

Also, the survival probability based on correlated clusters in the time evolution needs to be computed. Unlike the array evolution, the time evolution of correlated clusters performs from t_0 to T_{end}. The survival probability at t_k will be influenced by the previous time instant t_{k-1}, which satisfies $t_{k-1} = t_k - \Delta t$. This means that the survival probability will be updated continuously with respect to time. For dynamic clusters, the survival probability $P^{D,time}_{i,survival}(t_k)$ is given as

$$P^{D,time}_{i,survival}(t_k) = \sum_{i'=1}^{I(t_k)} \rho^{D}_{ii'}(t_k, f) P^{D,time}_{i,survival}(t_{k-1}). \tag{2.31}$$

Clearly, the survival probability of dynamic clusters $P^{D,time}_{i,survival}(t_k)$ can be obtained if the correlation coefficient matrix of dynamic clusters $\boldsymbol{\rho}^{D}(t, f)$ and the survival probability of dynamic clusters $P^{D,time}_{i,survival}(t_0)$ at the initial time instant t_0 are obtained. The correlation coefficient matrix $\boldsymbol{\rho}^{D}(t, f)$ has been given by (2.21). Similar to the array evolution, the survival probability of the ith dynamic cluster $P^{D,time}_{i,survival}(t_0)$ at the initial time instant t_0 can be expressed by

$$P^{D,time}_{i,survival}(t_0) =$$
$$\sum_{i'=1}^{I(t_0)} \left[x^{D}_{i'} \rho^{D}_{ii'}(t_0, f) e^{-\lambda^{D}_{D}\frac{(\Delta v^{D}_{T}+\Delta v^{D}_{R})\Delta t}{D^{S}_{c,D}}} + y^{D}_{i'} \rho^{D}_{ii'}(t_0, f)\left(1 - e^{-\lambda^{D}_{D}\frac{(\Delta v^{D}_{T}+\Delta v^{D}_{R})\Delta t}{D^{S}_{c,D}}}\right)\right].$$

Analogously, the survival probability of the jth static clusters $P^{S,time}_{j,survival}(t_k)$ can be computed by

$$P^{S,time}_{j,survival}(t_k) = \sum_{j'=1}^{J(t_k)} \rho^{S}_{jj'}(t_k, f) P^{S,time}_{j,survival}(t_{k-1}). \tag{2.32}$$

Additionally, the correlation coefficient matrix of static clusters $\boldsymbol{\rho}^S(t, f)$ has also been given by (2.21). The survival probability of the jth static clusters $P_{j,survival}^{S,time}(t_0)$ can be given as

$$P_{j,survival}^{S,time}(t_0) =$$

$$\sum_{j'=1}^{J(t_0)} \left[x_{j'}^S \rho_{jj'}^S(t_0, f) e^{-\lambda_D^S \frac{\left(\Delta v_T^S + \Delta v_R^S\right)\Delta t}{D_{c,S}^S}} + y_{j'}^S \rho_{jj'}^S(t_0, f) \left(1 - e^{-\lambda_D^S \frac{\left(\Delta v_T^S + \Delta v_R^S\right)\Delta t}{D_{c,S}^S}}\right)\right].$$

Similarly, the mean numbers of newly generated dynamic clusters and static clusters are further expressed as

$$E\left[I_{new}^D(t_{k+1})\right] = \frac{\lambda_B^D}{\lambda_D^D}\left(1 - \sum_{i=1}^{I(t_{k+1})} P_{i,survival}^{D,time}(t_{k+1} - t_k)/I(t_{k+1})\right) \tag{2.33}$$

$$E\left[J_{new}^S(t_{k+1})\right] = \frac{\lambda_B^S}{\lambda_D^S}\left(1 - \sum_{j=1}^{J(t_{k+1})} P_{j,survival}^{S,time}(t_{k+1} - t_k)/J(t_{k+1})\right). \tag{2.34}$$

2.3.2.4 Time-Array Evolution of Correlated Clusters

Based on the above mentioned three steps, the time-array evolution of dynamic and static correlated clusters can be characterized. After generating the dynamic and static correlated clusters, the first step is the initialization of correlated cluster sets. Then, novel birth-death processes based on dynamic/static correlated clusters in the array evolution and time evolution are presented in step 2 and step 3, respectively. Finally, in the time-array evolution, the survival probability of dynamic correlated clusters $P_{i,survival}^D(t_k)$ and static correlated clusters $P_{j,survival}^S(t_k)$ can be calculated as

$$P_{i,survival}^D(t_k) = P_{i,survival}^{T,D}(t_0) \cdot P_{i,survival}^{R,D}(t_0) \cdot P_{i,survival}^{D,time}(t_k) \tag{2.35}$$

$$P_{j,survival}^S(t_k) = P_{j,survival}^{T,S}(t_0) \cdot P_{j,survival}^{R,S}(t_0) \cdot P_{j,survival}^{S,time}(t_k). \tag{2.36}$$

By the aforementioned four steps, the dynamic/static correlated clusters have been obtained, and the consistency in birth and death between dynamic/static correlated clusters has also been modeled. As a result, the joint channel space-time-frequency non-stationarity of B5G/6G V2V mmWave massive MIMO channels has been properly modeled.

2.4 Simulations

2.4.1 *Statistical Properties Analysis*

In this section, important statistical properties of the proposed IS-GBSM, such as STF-CF and DPSD, will be derived and investigated sufficiently. Some interesting observations and conclusions will be presented.

2.4.1.1 Space-Time-Frequency Correlation Function

The time-variant transfer function $H_{qp}(t, f)$ can be derived by performing the Fourier transform of the CIR $h_{qp}(t, \tau)$ with respect to the delay, and can be written by

$$
\begin{aligned}
H_{qp}(t, f) = & \underbrace{\sqrt{\frac{\Omega}{\Omega+1}} h_{qp}^{LoS}(t) e^{-j2\pi f \tau^{LoS}(t)}}_{\text{LoS}} \\
& + \underbrace{\sqrt{\frac{\eta_1}{\Omega+1}} \sum_{i=1}^{I(t)} \sum_{n=1}^{N(t)} h_{qp,i,n}^{NLoS,D}(t) e^{-j2\pi f \left[\tau_i^D(t)+\tau_{i,n}^D(t)\right]}}_{\text{NLoS, Dynamic Clusters}} \\
& + \underbrace{\sqrt{\frac{\eta_2}{\Omega+1}} \sum_{j=1}^{J(t)} \sum_{m=1}^{M(t)} h_{qp,j,m}^{NLoS,S}(t) e^{-j2\pi f \left[\tau_j^S(t)+\tau_{j,m}^S(t)\right]}}_{\text{NLoS, Static Clusters}}.
\end{aligned}
\tag{2.37}
$$

Based on the time-variant transfer function $H_{qp}(t, f)$, the STF-CF can be calculated as

$$
\begin{aligned}
& \zeta_{qp,q'p'}(t, f; \Delta t, \Delta f, \delta_T, \delta_R) \\
& = E[H_{qp}^*(t, f) H_{q'p'}(t + \Delta t, f + \Delta f)]
\end{aligned}
\tag{2.38}
$$

where $(\cdot)^*$ designates the complex conjugate operation. Also, it is assumed that LoS components and NLoS components are independent of each other [29], and hence (2.38) can be expressed by (2.39), shown at the top of the next page,

$$
\begin{aligned}
& \zeta_{qp,q'p'}(t, f; \Delta t, \Delta f, \delta_T, \delta_R) = \zeta_{qp,q'p'}^{LoS}(t, f; \Delta t, \Delta f, \delta_T, \delta_R) + \\
& \zeta_{qp,q'p'}^{NLoS,D}(t, f; \Delta t, \Delta f, \delta_T, \delta_R) + \zeta_{qp,q'p'}^{NLoS,S}(t, f; \Delta t, \Delta f, \delta_T, \delta_R)
\end{aligned}
$$

where

$$\zeta^{LoS}_{qp,q'p'}(t,f;\Delta t,\Delta f,\delta_T,\delta_R)=\frac{\Omega}{\Omega+1}E\left[h^{LoS*}_{qp}(t,f)h^{LoS}_{q'p'}(t+\Delta t,f+\Delta f)e^{j\sigma^{LoS}}\right] \tag{2.39}$$

$$\zeta^{NLoS,D}_{qp,q'p'}(t,f;\Delta t,\Delta f,\delta_T,\delta_R)=\frac{\eta_1\bar{P}^D_{avg,survival}(t)}{\Omega+1}\times$$
$$E\left[\sum_{i=1}^{I(t)}\sum_{i'=1}^{I(t)}\sum_{n=1}^{N(t)}\sum_{n'=1}^{N'(t)}h^{NLoS,D*}_{qp,i,n}(t,f)h^{NLoS,D}_{q'p',i',n'}(t+\Delta t,f+\Delta f)e^{j\sigma^{NLoS,D}_{qp}}\right]$$

$$\zeta^{NLoS,S}_{qp,q'p'}(t,f;\Delta t,\Delta f,\delta_T,\delta_R)=\frac{\eta_2\bar{P}^S_{avg,survival}(t)}{\Omega+1}\times$$
$$E\left[\sum_{j=1}^{J(t)}\sum_{j'=1}^{J(t)}\sum_{m=1}^{M(t)}\sum_{m'=1}^{M'(t)}h^{NLoS,S*}_{qp,j,m}(t,f)h^{NLoS,S}_{q'p',j',m'}(t+\Delta t,f+\Delta f)e^{j\sigma^{NLoS,S}_{qp}}\right]$$

---

where $\sigma^{LoS}=2\pi[f\tau^{LoS}(t)-(f-\Delta f)\tau^{LoS}(t+\Delta t)]$, $\sigma_{qp}^{NLoS,D}=2\pi f[\tau_i^D(t)+\tau_{i,n}^D(t)]-2\pi(f+\Delta f)[\tau_{i'}^D(t+\Delta t)+\tau_{i',n'}^D(t+\Delta t)]$, $\sigma_{qp}^{NLoS,S}=2\pi f[\tau_j^S(t)+\tau_{j,m}^S(t)]-2\pi(f+\Delta f)[\tau_{j'}^S(t+\Delta t)+\tau_{j',m'}^S(t+\Delta t)]$. Clearly, STF-CFs of NLoS components should be scaled by the survival probability in the time-array evolution [19]. Unlike [19], in the proposed IS-GSBM, the survival probability of each cluster might be different due to the impact of correlated clusters. Here, STF-CFs of NLoS components are scaled by the average survival probabilities, which are given by $\bar{P}^D_{avg,survival}(t)=\sum_{i=1}^{I(t)}P^D_{i,survival}(t)/I(t)$ and $\bar{P}^S_{avg,survival}(t)=\sum_{j=1}^{J(t)}P^S_{j,survival}(t)/J(t)$.

Based on the expressions of STF-CF in (2.38), the spatial cross-correlation function (SCCF), time auto-correlation function (TACF), as well as frequency correlation function (FCF) can be further derived. By setting $\Delta t=\Delta f=0$, $p=p'(q=q')$, the STF-CF is simplified as SCCF. By setting $\Delta f=0$, $p=p',q=q'$, the STF-CF is simplified as TACF. By setting $\Delta t=0$, $p=p',q=q'$, the STF-CF is simplified as FCF.

2.4.1.2 Doppler Power Spectral Density

The DPSD with respect to Doppler frequency f_D can be obtained by performing the Fourier transfer of the TACF $\zeta_{qp,qp}(t;\Delta t)$. Based on the aforementioned analysis, the TACF $\zeta_{qp,qp}(t;\Delta t)$ can be expressed by

$$\zeta_{qp,qp}(t;\Delta t)=E[H^*_{qp}(t)H_{qp}(t+\Delta t)]. \tag{2.40}$$

In addition, the DPSD $\Gamma_{qp}(t;f_D)$ is given by

$$\Gamma_{qp}(t; f_D) = \int_{-\infty}^{+\infty} \zeta_{qp,qp}(t; \Delta t) e^{-j2\pi f_D \Delta t} \mathrm{d}(\Delta t). \tag{2.41}$$

It can be readily observed that the derived DPSD is time-variant, and thus the time-varying properties of V2V channels can be described.

2.4.2 Simulation Setting

Here, we will give the simulation results of the proposed IS-GBSM and analyze. Some valuable observations will be presented, and the utility of proposed IS-GBSM will be further verified. Note that the key parameters of dynamic clusters and static clusters are listed in Table 2.2, where $\mu_{A/E}^{T/R}$ and $\sigma_{A/E}^{T/R}$ denote the mean and standard deviation of the corresponding azimuth/elevation angles $\varphi_{i,n}^{T/R,D}(t)$ and $\gamma_{j,m}^{T/R,S}(t)$ of rays at Tx/Rx sides. To consider the ultra-high frequency band and ultra-wideband in mmWave communication scenarios, we set the carrier frequency $f_c = 28$ GHz [41] and bandwidth $BW = 2$ GHz. To support the massive MIMO scenario, the numbers of antenna elements at the Tx and Rx sides are $L_T = L_R = 32$. The initial distance between Tx and Rx is set as $D = 200$ m similar to [29]. To support the high and dual mobility of V2V communication scenarios, the velocities of Tx, Rx, and clusters are set as $v_T = 10$ m/s, $v_R = 10$ m/s, and $v_c = 10$ m/s, respectively. Furthermore, we have $\theta_T = \pi/3$, $\theta_R = 3\pi/4$, $\phi_T = \phi_R = \pi/4$, $\kappa = -8$ dB [29]. Unless otherwise specified, the aforementioned channel-related parameters are used in the simulations.

Table 2.2 Model-related parameters used in simulations

Parameter	Dynamic clusters	Static clusters
Space correlation coefficient $D_{c,D/S}^{s}$	10 m	100 m
Antenna correlation coefficient $D_{c,D/S}^{a}$	10 m	30 m
Mean value of $D_{i/j}^{T/R,D/S}$	5 m	15 m
μ_A^T, σ_A^T	$\pi/18, \pi/9$	$\pi/18, \pi/9$
μ_A^R, σ_A^R	$\pi, \pi/9$	$\pi, \pi/9$
μ_E^T, σ_E^T	$\pi/18, \pi/36$	$\pi/6, \pi/18$
μ_E^R, σ_E^R	$\pi/18, \pi/36$	$\pi/6, \pi/18$
High VTDs with initial ratio $\epsilon(t_0) = 3$		
Birth rate	$\lambda_B^D = 60$/m	$\lambda_B^S = 20$/m
Death rate	$\lambda_D^D = 4$/m	$\lambda_D^S = 4$/m
Low VTDs with initial ratio $\epsilon(t_0) = 1$		
Birth rate	$\lambda_B^D = 20$/m	$\lambda_B^S = 20$/m
Death rate	$\lambda_D^D = 4$/m	$\lambda_D^S = 4$/m

2.4.3 Simulation Results of the Proposed Model

Figure 2.5 presents the normalized absolute FCFs under high and low VTDs with frequencies $f = 27$ GHz and $f = 28$ GHz, which are in the communication bandwidth. It can be seen that the normalized absolute FCFs gradually decrease and then experience fluctuations when the frequency difference Δf increases, which is in agreement with the results in [42]. Furthermore, at the same central frequency, the proposed IS-GBSM has different FCFs at different frequencies. Therefore, FCFs not only depend on the frequency difference, but also on the frequency, demonstrating the frequency non-stationarity of the proposed IS-GBSM. Additionally, we notice that under different VTDs, the FCFs are significantly different. In comparison, FCFs with high VTDs are larger than FCFs with low VTDs. This phenomenon can be explained that, according to Table 2.2, the high VTD is a more cluttered environment, where there are more multipath components. This results in a larger delay spread, and thus FCFs with high VTDs are larger.

Figure 2.6 illustrates the normalized absolute TACFs with high and low VTDs at $t = 0.5$, $t = 1.5$, and $t = 3$ s. In Fig. 2.6, a decreasing trend of TACFs can be readily observed as the time difference increases, and TACFs are also dependable on time instant. Therefore, we can conclude that the time non-stationarity is modeled in the proposed IS-GBSM. Furthermore, compared to low VTDs, the simulated TACFs with high VTDs are lower. This is reasonable because, in high VTDs, there are lots of mobile vehicles in the environment, which leads to rapid changes in the

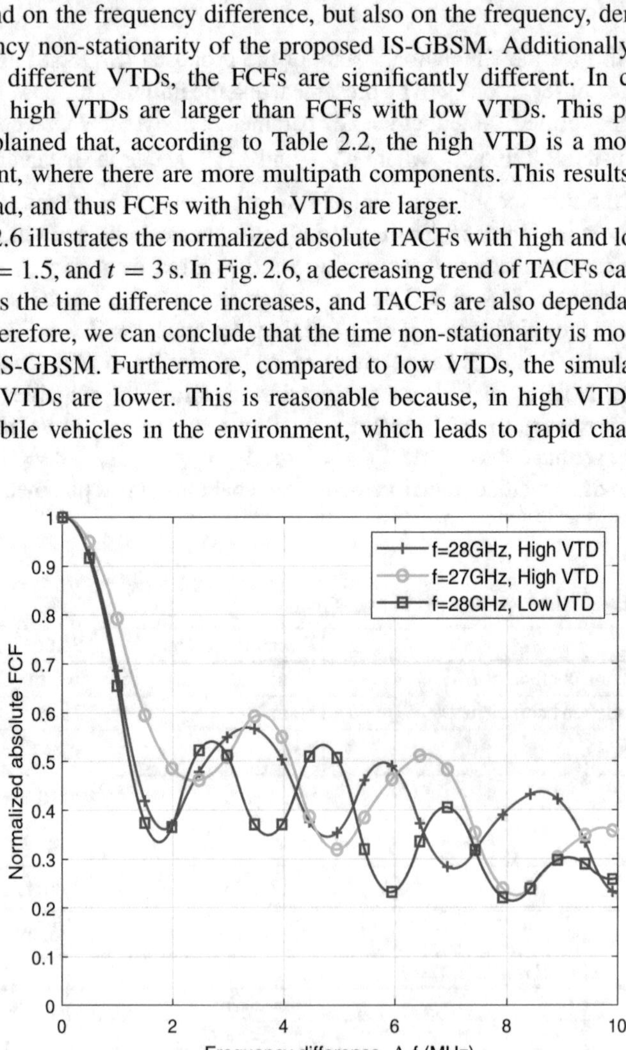

Fig. 2.5 Normalized absolute FCFs of the proposed IS-GBSM with different frequencies under different VTDs

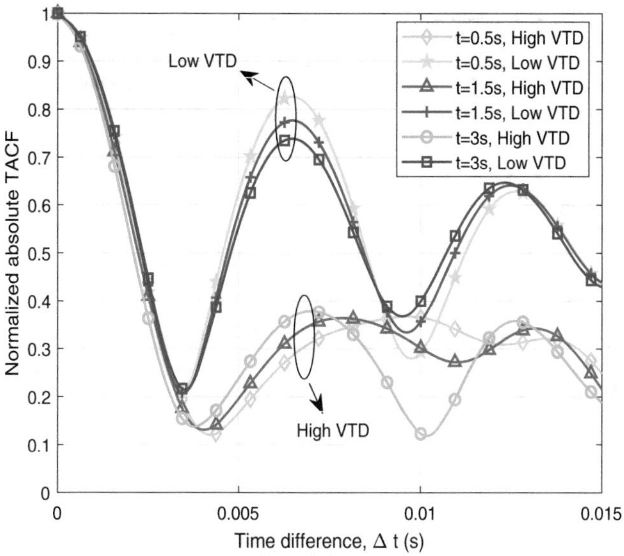

Fig. 2.6 Normalized absolute TACFs of the proposed IS-GBSM with different VTDs at different time instants

communication environment. In this case, the corresponding TACFs at different times are lower. Another phenomenon worth noting is that, the simulated TACFs are smoothly time-evolving. In comparison, as the time instant evolves from 0.5–3 s, the change of TACFs with high VTDs is more obvious than that of low VTDs. Again, this is due to the rapid change in the communication environment under high VTD scenarios.

Figure 2.7 shows the normalized absolute SCCFs at the Rx side with indices of reference antennas $s = 1$ and $s = 10$ under high and low VTDs. In Fig. 2.7, SCCFs of different values of s are different, and SCCFs further decrease as the antenna index difference enlarges. Therefore, the space non-stationarity is properly captured in the proposed IS-GBSM. In addition, compared to low VTDs, SCCFs with high VTDs are lower. The underlying physical reason is that the spatial diversity of channels increases as VTDs increase, leading to a decrease in spatial correlation. It is worth mentioning that similar conclusions on SCCFs can be found in [43].

2.4.4 Model Validation

Normalized DPSDs of the proposed IS-GBSM and measurement in [44] are compared in Fig. 2.8. The measurement campaign in [44] was carried out in the surface street area around the Georgia Tech campus and Interstate highway in the Midtown Atlanta metropolitan area. From Fig. 3 in [44], we can observe that there

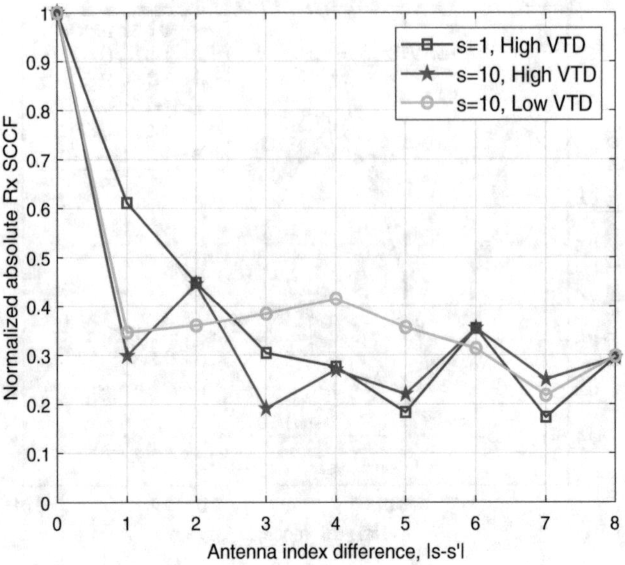

Fig. 2.7 Normalized absolute SCCFs of the proposed IS-GBSM at the Rx side with different indices of reference antennas s under different VTDs

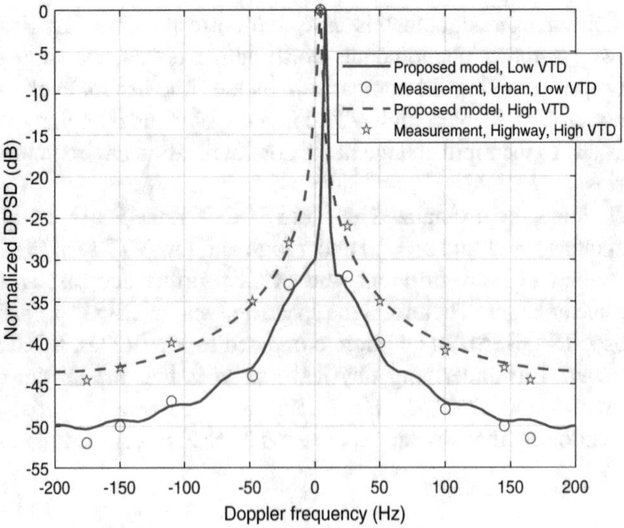

Fig. 2.8 Normalized DPSDs of the proposed IS-GBSM and the measurement in [44] under different VTDs

are few moving vehicles, which can be regarded as a low VTD scenario. In contrast, from Fig. 4 in [44], it is clear that there are many moving vehicles on the highway, which can be regarded as a high VTD scenario. For a fair comparison, parameters of the proposed IS-GBSM are set based on the measurement [44], i.e., $f_c = 2.435$ GHz, $\delta_T = \delta_R = 2.943\lambda$, and $\theta_T = \theta_R = \phi_T = \phi_R = 0°$. In the low VTD, $v_T = v_R = 11$ m/s and $D = 300$ m. In the high VTD, $v_T = v_R = 22$ m/s and $D = 180$ m. In Fig. 2.8, it can be readily observed that the VTD has a significant impact on the distribution of DPSDs. Specifically, compared to low VTDs, DPSDs with high VTDs exhibit more flatly distributed. It is because the received power in high VTDs tends to come from dynamic cars over all directions. On the contrary, the received power in low VTDs mainly concentrates on several Doppler frequencies, leading to steep distributions of DPSDs. Note that the simulated DPSDs match well with measurements, verifying the utility of proposed IS-GSBM.

Figure 2.9 compares the normalized absolute TACFs of the proposed IS-GBSM with the measurement in [45]. The measurement campaign of MIMO vehicular channels in [45] was performed at 5.8 GHz, where the Rx antennas on the vehicle were away from the fixed Tx at a speed of 60 km/h. For a fair comparison, parameters of the proposed IS-GBSM are set as $f_c = 5.8$ GHz, $L_T = L_R = 4$, $\delta_T = \delta_R = 2\lambda$, $v_T = 0$ m/s, and $v_R = 16.667$ m/s. Based on [45], since there are few vehicles moving in the measured environment, it can be assumed that it is a low VTD scenario. Note that, high VTD scenarios were not considered in [45]. Therefore, we only compare the simulated TACF and measurement data under low VTDs. From Fig. 2.9, a good agreement is achieved between the simulated TACF and measurement, validating the accuracy of the proposed IS-GBSM. Additionally, it is clear that channels with high VTDs exhibit a lower temporal correlation than channels with low VTDs. This is consistent with the conclusion drawn in Fig. 2.6.

In Fig. 2.10, the CDFs of delayed component amplitude in the proposed IS-GSBM and measurement [46] are compared. Note that the delayed component has an excess delay of 7 ns. Furthermore, measurement campaigns are carried out on the university of Vigo Campus and the bypass road, which can be reasonably assumed to be a high VTD scenario. Based on the measurement [46], corresponding model-related parameters of the proposed IS-GSBM are set as $f_c = 38$ GHz, $v_T = v_R = 19.44$ m/s, and $D = 123$ m. It can be observed from Fig. 2.10 that the amplitude relative to mean is mainly concentrated between −6 and 6 dB. Also, a good match between simulation results and measurements is achieved, which confirms the accuracy of proposed IS-GSBM.

2.4.5 Model Application

Based on the simulation results shown in this Section, it is clear that the space-time-frequency non-stationarity has been modeled in the proposed IS-GBSM. By modeling the channel space-time-frequency non-stationarity, the proposed IS-GBSM and the understanding of channel non-stationarity can provide some ideas and guidance for the design of B5G/6G vehicular communication systems. Then, the proposed IS-GBSM can be used as a simulation and validation platform, providing some guidance for the research of B5G/6G vehicular communication

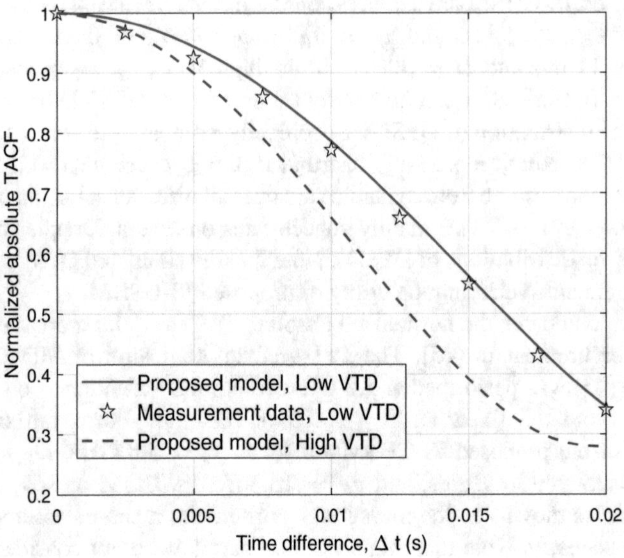

Fig. 2.9 Normalized TACFs of the proposed IS-GBSM and the measurement in [45]

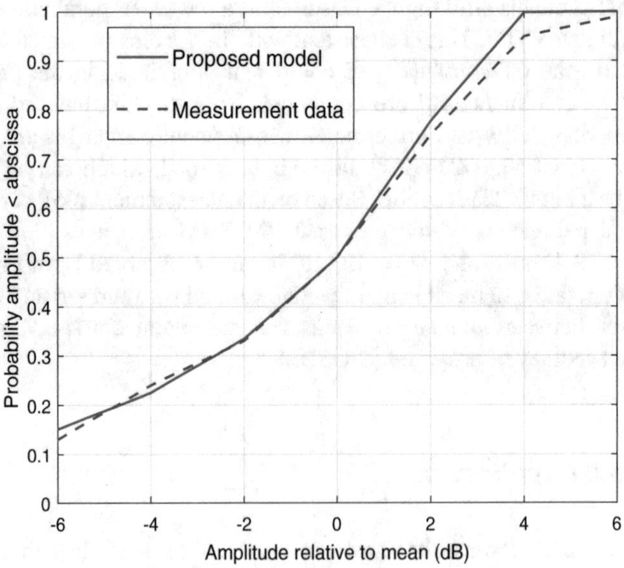

Fig. 2.10 CDFs of the amplitude of delayed component in the proposed IS-GSBM and the measurement in [46]

system algorithms. Furthermore, simulation results have demonstrated that the VTD has a great impact on channel statistics. Compared to low VTDs, B5G/6G V2V channels with high VTDs explicitly exhibit higher FCFs, lower TACFs, lower SCCFs. These interesting observations are instructive and helpful for the design of pilot, the determination of the transmission rate, and the diversity and multiplexing gain of massive MIMO. As a result, the proposed V2V massive MIMO mmWave IS-GBSM can efficiently guide the B5G/6G vehicular communication system design and performance evaluation in practice, which can support the development of the future ITS in practice.

In addition to the influence of space-time-frequency non-stationarity on the point-to-point physical layer (PHY) design, as mentioned in [6], the impact of non-stationarity and VTDs also needs to be modeled in vehicular channel models even for research and simulations at the network level. As a result, the proposed IS-GBSM can also support the network level algorithm research and design for the vehicular communication system, which is a key component of the ITS.

Finally, it is clear that some important ITS applications, e.g., multi-vehicle cooperative perception, vehicle positioning, and vehicle tracking, will make use of mmWave Massive MIMO channels. By considering the impact of mmWave and massive MIMO technologies on channel modeling, the proposed IS-GBSM can also support these ITS applications, which can significantly improve safety, environmental footprint, and transportation efficiency.

2.5 Discussions and Summary

In this chapter, a 3D space-time-frequency non-stationary IS-GBSM for B5G/6G V2V mmWave massive MIMO channels has been proposed. By distinguishing clusters into dynamic clusters and static clusters, it is the first channel model that can investigate the impact of VTDs on channel statistics in B5G/6G V2V mmWave massive MIMO scenarios. Rays within dynamic clusters and static clusters have been resolved to support mmWave channels with high delay resolution. In addition, a novel method based on dynamic/static correlated clusters has been developed to mimic the channel space-time-frequency non-stationarity. Based on the proposed IS-GBSM, comprehensive channel statistical properties have been derived and sufficiently analyzed. Simulation results have demonstrated that the proposed IS-GBSM can model the space-time-frequency non-stationarity and the VTD has a great impact on channel statistics. Compared to low VTDs, B5G/6G V2V channels with high VTDs explicitly exhibit higher FCFs, lower TACFs, lower SCCFs, and more flatly distributed DPSDs. Note that the above mentioned observations can provide useful guidance for the design of B5G/6G V2V systems. Finally, the simulated DPSD, TACF, and CDF of delayed component amplitude can closely agree with the measurement, which has verified the utility of the proposed IS-

GBSM. In the future study, we aim to carry out measurements in realistic V2V scenarios to further validate the utility of the proposed non-stationary IS-GBSM for massive MIMO mmWave V2V channels.

References

1. W. Viriyasitavat, M. Boban, H. Tsai, A. Vasilakos, Vehicular communications: survey and challenges of channel and propagation models. IEEE Veh. Technol. Mag. **10**(2), 55–66 (2015)
2. I. Sen, D.W. Matolak, Vehicle-vehicle channel models for the 5-GHz band. IEEE Trans. Intell. Transp. Syst. **9**(2), 235–245 (2008)
3. C.X. Wang, X. Cheng, D.I. Laurenson, Vehicle-to-vehicle channel modeling and measurements: recent advances and future challenges. IEEE Commun. Mag. **47**(11), 96–103 (2009)
4. R. He, B. Ai, G.L. Stuber, G. Wang, Z. Zhong, Geometrical-based modeling for millimeter-wave MIMO mobile-to-mobile channels. IEEE Trans. Veh. Tech. **67**(4), 2848–2863 (2018)
5. R. He, et al., Propagation channels of 5G millimeter-wave vehicle-tovehicle communications: recent advances and future challenges. IEEE Veh. Technol. Mag. **15**(1), 16–26 (2020)
6. X. Cheng, Z. Huang, S. Chen, 6G vehicular communication channel measurement, modeling, and application. IET Commun. **14**(19), 3303–3311 (2020)
7. X. Cheng, R. Zhang, L. Yang, *5G-Enabled Vehicular Communications and Networking*, 1st edn. (Springer, Cham, 2019)
8. X. Cheng, C. Chen, W. Zhang, Y. Yang, 5G-enabled cooperative intelligent vehicular (5GenCIV) framework: when Benz meets Marconi. IEEE Intell. Syst. **32**(3), 53–59 (2017)
9. X. Cheng, Q. Yao, M. Wen, C.-X. Wang, L. Song, B. Jiao, Wideband channel modeling and intercarrier interference cancellation for vehicle-to-vehicle communication systems. IEEE J. Sel. Areas Commun. **31**(9), 434–448 (2013)
10. J. Maurer, T. Fgen, W. Wiesbeck, A ray-optical channel model for vehicle-to-vehicle communication, in *Proceedings in Physics: Fields, Networks, Computational Methods, and Systems in Modern Electrodynamics*, ed. by P. Russer, M. Mongiardo (Springer, Berlin, 2004), pp. 243–254
11. L. Reichardt, T. Fugen, T. Zwick, Influence of antennas placement on car to car communications channel, in *Proceedings of ECAP'09, Berlin* (2009), pp. 630–634
12. W. Wiesbeck, S. Knorzer, Characteristics of the mobile channel for high velocities, in *Proceedings of ICEAA'07, Torino* (2007), pp. 116–120
13. G. Acosta-Marum, M.A. Ingram, Six time- and frequency- selective empirical channel models for vehicular wireless LANs. IEEE Veh. Technol. Mag. **2**(4), 4–11 (2007)
14. Z. Huang, X. Zhang, X. Cheng, Non-geometrical stochastic model for non-stationary wideband vehicular communication channels. IET Commun. **14**(1), 54–62 (2020)
15. Z. Huang, X. Cheng, A general 3D space-time-frequency non-stationary model for 6G channels. IEEE Trans. Wireless Commun. **20**(1), 535–548 (2021)
16. X. Cheng, R. Zhang, L. Yang, Wireless towards the era of intelligent vehicles. IEEE Internet. Things. J. **6**(1), 188–202 (2019)
17. A.K. Akki, F. Haber, A statistical model of mobile-to-mobile land communication channel. IEEE Trans. Veh. Technol. **35**(1), 2–7 (1986)
18. X. Cheng, C.-X. Wang, D.I. Laurenson, S. Salous, A.V. Vasilakos, An adaptive geometry-based stochastic model for non-isotropic MIMO mobile-to-mobile channels. IEEE Trans. Wireless Commun. **8**(9), 4824–4835 (2009)
19. A.G. Zajić, G.L. Stüber, Three-dimensional modeling, simulation, and capacity analysis of space–time correlated mobile-to-mobile channels. IEEE Trans. Veh. Technol. **57**(4), 2042–2054 (2008)

20. A.G. Zajić, G.L. Stüber, Three-dimensional modeling and simulation of wideband MIMO mobile-to-mobile channels. IEEE Trans. Wireless Commun. **8**(3), 1260–1275 (2009)
21. X. Gao, F. Tufvesson, O. Edfors, Massive MIMO channels — measurements and models, in *Proceddings of 2013 Asilomar Conference on Signals, Systems and Computers, Pacific Grove, CA* (2013), pp. 280–284
22. H. Jiang, C. Chen, J. Zhou, J. Dang, L. Wu, A general 3D non-stationary wideband twin-cluster channel model for 5G V2V tunnel communication environments. IEEE Access **7**, 137744–137751 (2019)
23. L. Bai, Z. Huang, H. Du, X. Cheng, A 3-D non-stationary wideband V2V GBSM with UPAs for massive MIMO wireless communication systems. IEEE Internet. Things. J. (to be published). https://doi.org/10.1109/JIOT.2021.3081816
24. C.-X. Wang, J. Huang, H. Wang, X. Gao, X. You, Y. Hao, 6G wireless channel measurements and models: trends and challenges. IEEE Veh. Technol. Mag. **15**(4), 22–32 (2020)
25. B.M. Eldowek, et al. 3D non-stationary vehicle-to-vehicle MIMO channel model for 5G millimeter-wave communications. Digit. Signal Process. **95**, 102580 (2019)
26. N. Czink, F. Kaltenberger, Y. Zhou, L. Bernadó, T. Zemen, X. Yin, Low-complexity geometry-based modeling of diffuse scattering, in *Proceedings of the Fourth European Conference on Antennas and Propagation (EuCAP), Barcelona* (2010), pp. 1–4
27. A. Ghazal, et al., A non-stationary IMT-advanced MIMO channel model for high-mobility wireless communication systems. IEEE Trans. Wireless Commun. **16**(4), 2057–2068 (2017)
28. O. Renaudin, V. Kolmonen, P. Vainikainen, C. Oestges, Wideband measurement-based modeling of inter-vehicle channels in the 5-GHz band. IEEE Trans. Veh. Technol. **62**(8), 3531–3540 (2013)
29. S. Wu, C.-X. Wang, H. Aggoune, M.M. Alwakeel, X.-H. You, A general 3-D non-stationary 5G wireless channel model. IEEE Trans. Commun. **66**(7), 3065–3078 (2018)
30. L. Bai, Z. Huang, Y. Li, X. Cheng, A 3D cluster-based channel model for 5G and beyond vehicle-to-vehicle massive MIMO channels. IEEE Trans. Veh. Technol. **70**(9), 8401–8414 (2021)
31. H. Hofstertter, A.-F. Molisch, N. Czink, A twin-cluster MIMO channel model, in *Proceedings of EuCAP, Nice* (2006), pp. 1–8
32. G.L. Stüber, *Principles of Mobile Communications*, 2nd edn. (Norwell, Kluwer, 2011)
33. M. Lehsaini, M.B. Benmahdi, An improved K-means cluster-based routing scheme for wireless sensor networks, in *Proceedings of ISPS, Algiers* (2018), pp. 1–6
34. P. Bholowalia, A. Kumar, EBK-means: a clustering technique based on Elbow method and K-means in WSN. Int. J. Comput. Appl. **105**(9), 17–24 (2014)
35. D. Marutho, S.H. Handaka, E. Wijaya, Muljono, The determination of cluster number at k-mean using elbow method and purity evaluation on headline news, in *Proceedings of ISEMANTIC, Semarang* (2018), pp. 533–538
36. R.C. Qiu, I.-T. Liu, Multipath resolving with frequency dependence for wide-band wireless channel modeling. IEEE Trans. Veh. Technol. **48**(1), 273–285 (1999)
37. R.C. Qiu, A study of the ultra-wideband wireless propagation channel and optimum UWB receiver design. IEEE J. Sel. Areas Commun. **20**(9), 1628–1637 (2002)
38. M.R. Akdenizet, et al. Millimeter wave channel modeling and cellular capacity evaluation. IEEE J. Sel. Areas Commun. **32**(6), 1164–1179 (2014)
39. Study on channel model for frequencies from 0.5 to 100 GHz, version 14.2.0. Document 3GPP T.R. 38.901 (2017)
40. A. Papoulis, S.U. Pillai, *Probability, Random variables, Stochastic Processes*, 4nd edn. (McGraw-Hill, New York, 2002)
41. J. Park, J. Lee, J. Liang, K. Kim, K. Lee, M. Kim, Millimeter wave vehicular blockage characteristics based on 28 GHz measurements, in *Proceedings of 2017 IEEE VTC-Fall, Toronto* (2017), pp. 1–5
42. S. Wu, C.-X. Wang, H. Haas, E.-H.M. Aggoune, M.M. Alwakeel, B. Ai, A non-stationary wideband channel model for massive MIMO communication systems. IEEE Trans. Wireless Commun. **14**(3), 1434–1446 (2015)

43. Y. Li, X. Cheng, N. Zhang, Deterministic and stochastic simulators for non-isotropic V2V-MIMO wideband channels. China Commun. **15**(7), 18–29 (2018)
44. A.G. Zajić, G.L. Stüber, T.G. Pratt, S.T. Nguyen, Wideband MIMO mobile-to-mobile channels: geometry-based statistical modeling with experimental verification. IEEE Trans. Veh. Technol. **58**(2), 517–534 (2009)
45. A. Fayziyev, M. Pätzold, E. Masson, Y. Cocheril, M. Berbineau, A measurement-based channel model for vehicular communications in tunnels, in *Proceedings of IEEE WCNC'14, Istanbul* (2014), pp. 1–5
46. M. García, M. Táboas, E. Cid, Millimeter wave radio channel characterization for 5G vehicle-to-vehicle communications. Measurement **95**, 223–229 (2018)

Chapter 3
Millimeter-Wave Vehicular Channel Estimation

Abstract This chapter works on designing an efficient channel estimator for hybrid mmWave massive multiple-input multiple-output (mMIMO) systems. The proposed doubly-sparse approach relies on a judiciously designed training pattern to decouple the convoluted channel. By doing so, it becomes convenient to exploit the under-investigated channel sparsity in the delay domain together with the well-known beamspace sparsity. Furthermore, dedicated probing strategies are accordingly developed to ensure compatibility with the hybrid structure while utilizing double sparsity. Compared with existing alternatives, the proposed mmWave channel estimator works exceptionally in doubly-selective (frequency-time) channels and can hugely reduce the training overhead, storage demand, and computational complexity thanks to the exploitation of double (delay-beamspace) sparsity.

Keywords mmWave · Hybrid structure · Massive multiple-input multiple-output · Channel estimation · Double selectivity · Double sparsity · Frequency-time · Delay-beamspace

3.1 Background

3.1.1 Necessity of Doubly-Selective Channel Estimator

The foundation of wireless transceiver design lies in accurate channel state information (CSI) [1–3]. However, channel estimation for mmWave mMIMO faces unprecedented challenges, compared to the conventional centimeter-wave (cmWave) MIMO. The first challenge arises from the hybrid structure, which requires the high-dimensional channel matrix to be recovered via very few RF chains [4–6]. As the latter essentially determines how many effective training symbols can be transmitted simultaneously, significant amount of time will be needed in order to transmit sufficient training symbols. The problem becomes even more challenging in mobile scenarios because the propagation environments may have already changed before all training symbols are sent out.

X. Cheng et al., *mmWave Massive MIMO Vehicular Communications*,
Wireless Networks, https://doi.org/10.1007/978-3-030-97508-1_3

Our channel modeling works in conjunction with many other relevant literature have shown that mmWave channels exhibit limited scattering [7]. This property renders a unique sparsity situation in beamspace under mMIMO. As a result, estimating the channel matrix element by element may not be necessary. Instead, compressed sensing (CS) theory can come into play to reduce the training overhead while ensuring high accuracy [8]. Following this idea, some representative works include [9–12]. To the best of our knowledge, all these channel estimation works for hybrid mmWave mMIMO consider at most one selectivity, that is, either the frequency selectivity or the time selectivity. Clearly, to meet general application scenarios, channel estimation should account for both the frequency selectivity and the time selectivity.

3.1.2 Design Objectives and Proposed Approaches

Intuitively, one needs to take advantage of the under-exploited delay-domain sparsity in combination with the well-known beamspace sparsity to achieve significant reduction in training overhead and computational complexity, As a matter of fact, the idea of using either the delay-domain sparsity or the double sparsity can be also found in previous literature [13–15]. However, they are not specifically designed for hybrid mMIMO and the studied channels therein have not taken time selectivity into account. In fact, once the time selectivity is involved, the exploitation of delay-domain sparsity or the beamspace sparsity becomes a touchy problem. Besides, the mmWave hybrid structure further aggravates the difficulty since the design flexibility is significantly limited by the hardware constraints.

To overcome these obstacles, a novel doubly-sparse approach is hereby proposed for mmWave mMIMO channel estimation. The so-called DSDS channel estimator comprises the following key procedures.

- To deal with the delay-domain sparsity, a special training pattern is judiciously designed to facilitate channel tap separation. In consequence, only a small portion of channel taps will be identified based on the energy detector, regardless of the Doppler effects.
- To deal with the beamspace sparsity, an enhanced orthogonal matching pursuit (OMP) algorithm, namely, A-BOMP, is proposed to for beam estimation. A-BOMP can adjust basis matching and residue update in accordance with the maximum Doppler such that it maintains strong robustness against Doppler effects.
- To estimate the channel tap amplitudes and Doppler, a steered-probing polling method is applied given the estimated beam direction, which can yield high accuracy at a small training cost.

3.2 System and Channel Models

3.2.1 System Model

We consider a mmWave mMIMO system where N_t and N_r antennas are employed at the transmitter (Tx) and receiver (Rx), respectively. Since no channel reciprocity assumption is made, we therefore assume the estimation is implemented at Rx. Similar to [16], a fully-connected hybrid structure is studied here. In this structure, there are much fewer RF chains than the antennas, with their connection established via a digitally controlled analog phase shifter (APS) network. Suppose each APS component has a resolution of b bits. All the adjustable angles can be represented as

$$\mathcal{B} = \left\{0, 2\pi/2^b, \cdots, 2\pi(2^b - 1)/2^b\right\} \tag{3.1}$$

with $|\mathcal{B}| = 2^b$. Accordingly, the angular quantization function is expressed as

$$\mathcal{Q}(x) = \mathcal{B}(i^*), i^* = \underset{i}{\text{argmin}} \mod \left(x - \mathcal{B}(i), 2\pi\right). \tag{3.2}$$

Without loss of generality, we assume that the transceivers each employ a single RF chain as shown in Fig. 3.1. The proposed methods can be readily generalized to cope with arbitrary number of RF chains.

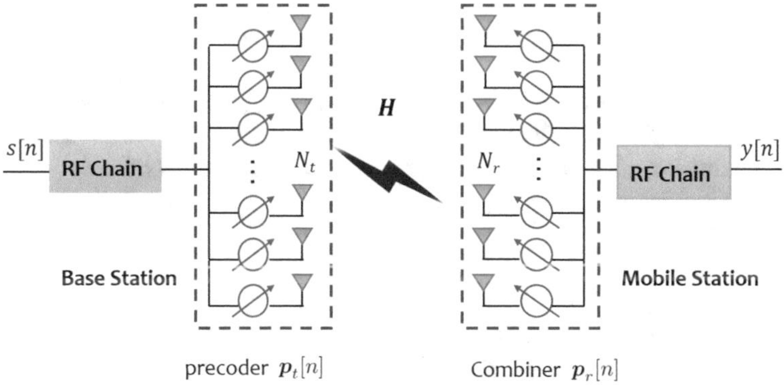

Fig. 3.1 The system model of hybrid mmWave mMIMO transceivers with one RF chain deployed at either end

3.2.2 Channel Models

Stemming from the prevalent geometric-based mmWave channel modeling, a generalized geometric channel model [17] is adopted to describe the channel matrix per realization. Let N_c be the maximum number of delay taps. The tap-d channel $(0 \leq d < N_c)$ sampled at time instant n can be given by

$$\boldsymbol{H}_d(n)=\sum_{p=1}^{P}\sqrt{\frac{N_t N_r}{P}}\alpha_p h(dT_s-\tau_p)\boldsymbol{a}_r(\theta_p)\boldsymbol{a}_t^*(\phi_p)e^{j\omega_p n}. \tag{3.3}$$

In the above, $\alpha_p \sim \mathcal{CN}(0, 1)$ is the complex gain of the p-th path; $h(\cdot)$ is the pulse shaping filter response; τ_p is the propagation delay of the p-th path that obeys a uniform distribution on $[0, (N_c - 1)T_s)$; θ_p and ϕ_p are the angle of arrival (AoA) and angle of departure (AoD), both uniformly distributed on $[0, 2\pi)$. Let the system carrier frequency be f_c, the velocity of light be c_v, and the maximum relative velocity be v_m. The normalized Doppler shift can accordingly be calculated as $\omega_p = 2\pi f_c v_m T_s \sin(\theta_p)/c_v$. For simplicity, we refine an the following function:

$$\boldsymbol{f}_N(y) = \frac{1}{\sqrt{N}}\left[1, e^{j2\pi y}, \cdots, e^{j2\pi(N-1)y}\right]'. \tag{3.4}$$

Then the array response for a half-wavelength uniform linear array (ULA) can be given by $\boldsymbol{a}_t(\phi) = \boldsymbol{f}_{N_t}\big(\sin(\phi)/2\big)$.

The geometric model can be further written into the following compact form:

$$\boldsymbol{H}_d(n) = \boldsymbol{A}_R\mathbf{diag}\big(\boldsymbol{g}_d(n)\big)\boldsymbol{A}_T^*. \tag{3.5}$$

Take the transmitter as an example. $\boldsymbol{A}_T = \left[\boldsymbol{a}_t(\phi_1), \boldsymbol{a}_t(\phi_2), \cdots, \boldsymbol{a}_t(\phi_P)\right] \in \mathcal{C}^{N_t \times P}$ is a steering matrix that remains unchanged during the entire channel estimation stage. The time-varying effects are incorporated in $\boldsymbol{g}_d(n)$ given by

$$\sqrt{\frac{N_t N_r}{P}}\left[\alpha_1 h(dT_s - \tau_1)e^{j\omega_1 n}, \cdots, \alpha_P h(dT_s - \tau_P)e^{j\omega_P n}\right]'. \tag{3.6}$$

In Eq. (3.5), $\boldsymbol{A}_T, \boldsymbol{A}_R$ as well as $\boldsymbol{g}_d(n)$ are all associated with the physical channel taps, which may not be resolvable due to the finite space-time resolution. To seek an equally general but more practical expression, we first construct the Tx-end and Rx-end angular dictionary matrices as

$$\boldsymbol{D}_t = \left[\boldsymbol{f}_{N_t}(0), \boldsymbol{f}_{N_t}(1/G_t), \cdots, \boldsymbol{f}_{N_t}\big((G_t-1)/G_t\big)\right] \tag{3.7a}$$

$$\boldsymbol{D}_r = \left[\boldsymbol{f}_{N_r}(0), \boldsymbol{f}_{N_r}(1/G_r), \cdots, \boldsymbol{f}_{N_r}\big((G_r-1)/G_r\big)\right] \tag{3.7b}$$

where G_t (G_r) represents the dictionary size . $\boldsymbol{D}_t$ contains all steering vectors ranging from $[0, 2\pi]$ with resolution $2\pi/G_t$. The resolution becomes zero if G_t approaches infinity, thus resulting in a continuous dictionary. In practice, this value has to be finite, and most works show that setting G_t as $2 \sim 4$ times the array size suffices to separate AoAs/AoDs. With the dictionary matrices, the channel representation in Eq. (3.5) can be re-expressed as

$$\boldsymbol{H}_d(n) = \boldsymbol{A}_R \mathbf{diag}\big(\boldsymbol{g}_d(n)\big)\boldsymbol{A}_T^* = \boldsymbol{D}_r \overline{\boldsymbol{H}}_d(n)\boldsymbol{D}_t^*. \tag{3.8}$$

Under the mMIMO setup, P propagation paths result in P dominant non-zero entries in $\overline{\boldsymbol{H}}_d$. As $\boldsymbol{D}_r$ and $\boldsymbol{D}_t$ are irrelevant to $\boldsymbol{H}_d$, $\overline{\boldsymbol{H}}_d$ essentially gathers the entire channel information that was originally contained by $\boldsymbol{A}_T$, $\boldsymbol{A}_R$ and $\boldsymbol{g}_d(n)$. Specifically, by omitting the time instant and assuming on-grid AoA/AoD pairs, $\forall p \in [1, P]$, $n_p = \frac{\phi_p}{2\pi/G_t}$, $m_p = \frac{\theta_p}{2\pi/G_t}$, $\overline{\boldsymbol{H}}_d(m_p, n_p) = \boldsymbol{g}_d[p]$. From this sense, $\overline{\boldsymbol{H}}_d$ can be interpreted as the channel representation in beamspace.

In the rest chapter, we will rely on the above beamspace representation to obtain beam direction (AoA & AoD), beam amplitude, as well as the associated Doppler shift.

3.2.3 Input-Output Relationship

Let $s(n)$ be the training symbol at instant-n. At the Tx, $s(n)$ is first processed at the APS network, and the transmitted signal is $\boldsymbol{x}(n) = \boldsymbol{p}_t(n)s(n) \in \mathcal{C}^{N_t \times 1}$. Since each APS component can only adjust the phase, the probing vector $\boldsymbol{p}_t(n)$ bears the form as

$$\boldsymbol{p}_t(n) = \sqrt{1/N_t}\big[e^{j\alpha_1(n)}, e^{j\alpha_2(n)}, \cdots, e^{j\alpha_{N_t}(n)}\big]' \tag{3.9}$$

with $\alpha_i(n) \in \mathcal{B}$, $\forall i \in [1, N_t]$.

After channel propagation, the received signal is

$$\boldsymbol{r}(n) = \sum_{d=0}^{N_c-1} \boldsymbol{H}_d(n)\boldsymbol{x}(n-d) + \boldsymbol{\eta}(n) \tag{3.10}$$

which is the convolution of multiple time-varying channel taps. $\boldsymbol{\eta}(n) \sim \mathcal{CN}(\mathbf{0}, \sigma^2 \boldsymbol{I}_{N_r})$ is the receiver noise vector. $\boldsymbol{r}(n)$ then goes through the Rx-end APS network, whose function is described by an $N_r \times 1$ probing vector $\boldsymbol{p}_R(n)$, so the received sample after APS becomes

$$y(n) = \sum_{d=0}^{N_c-1} \boldsymbol{p}_r^*(n)\boldsymbol{H}_d(n)\boldsymbol{p}_t(n-d)s(n-d) + \xi(n). \tag{3.11}$$

where $\xi(n) = \boldsymbol{p}_r^*(n)\boldsymbol{\eta}(n) \sim \mathcal{CN}(0, \sigma^2)$ remains white. Let $\overline{\boldsymbol{p}}_t(n) = \boldsymbol{D}_t \boldsymbol{p}_t(n)$ and $\overline{\boldsymbol{p}}_r(n) = \boldsymbol{D}_r \boldsymbol{p}_r(n)$. Based on the beamspace representation in Eq. (3.8), we have

$$y(n) = \sum_{d=0}^{N_c-1} \overline{\boldsymbol{p}}_r^*(n)\overline{\boldsymbol{H}}_d(n)\overline{\boldsymbol{p}}_t(n-d)s(n-d) + \xi(n). \tag{3.12}$$

Without loss of generality, we consider the general I-O relationship for the first frame only unless otherwise specified. The length-N_f training frame is simply denoted as $[s(0), s(1), \cdots, s(N_f-1)]$, and its specific form will be explained later. By concatenating all received samples, we write the I-O relationship in matrix form shown in Eq. (3.13) at the top of next page,

$$\begin{aligned} \boldsymbol{y} &= [y(0), y(1), \cdots, y(N_f-1)]' \\ &= \overline{\boldsymbol{P}}_r^* \begin{bmatrix} \overline{\boldsymbol{H}}_0(0) & \boldsymbol{0} & \boldsymbol{0} & \cdots & \boldsymbol{0} \\ \vdots & \overline{\boldsymbol{H}}_0(1) & \boldsymbol{0} & \cdots & \boldsymbol{0} \\ \overline{\boldsymbol{H}}_{N_c-1}(N_c-1) & \cdots & \ddots & \cdots & \vdots \\ \vdots & \ddots & \cdots & \ddots & \boldsymbol{0} \\ \boldsymbol{0} & \vdots & \overline{\boldsymbol{H}}_{N_c-1}(N_f-1) & \cdots & \overline{\boldsymbol{H}}_0(N_f-1) \end{bmatrix} \times \\ &\quad \overline{\boldsymbol{P}}_t \begin{bmatrix} s(0) \\ \vdots \\ s(N_c-1) \\ \vdots \\ s(N_f-1) \end{bmatrix} + \boldsymbol{\xi}. \end{aligned} \tag{3.13}$$

with $\overline{\boldsymbol{P}}_r^* = \begin{bmatrix} \overline{\boldsymbol{p}}_r^*(0) & \boldsymbol{0} & \cdots & \boldsymbol{0} \\ \boldsymbol{0} & \overline{\boldsymbol{p}}_r^*(1) & \cdots & \boldsymbol{0} \\ \vdots & \vdots & \ddots & \boldsymbol{0} \\ \boldsymbol{0} & \boldsymbol{0} & \cdots & \overline{\boldsymbol{p}}_r^*(N_f-1) \end{bmatrix}$ and $\overline{\boldsymbol{P}}_t = \begin{bmatrix} \overline{\boldsymbol{p}}_t(0) & \boldsymbol{0} & \cdots & \boldsymbol{0} \\ \boldsymbol{0} & \overline{\boldsymbol{p}}_t(1) & \cdots & \boldsymbol{0} \\ \vdots & \vdots & \ddots & \boldsymbol{0} \\ \boldsymbol{0} & \boldsymbol{0} & \cdots & \overline{\boldsymbol{p}}_t(N_f-1) \end{bmatrix}$.

3.3 Channel Estimation via Exploiting Double Sparsity

As mentioned earlier, mmWave channels exhibit sparsity in beamspace. Apart from this well-known sparsity, this section will further reveal that mmWave channels exhibit sparsity in the delay domain as well. We start by explaining why existing approaches fail to exploit the delay-domain sparsity. Then we elaborate on the benefits if properly utilizing this largely overlooked sparsity.

3.3.1 Proposed Training Pattern

The currently adopted training pattern has the following form:

$$\big[s_0, s_1, s_2, \cdots, s_{N-1}, \underbrace{0, \cdots, 0}_{N_c}\big]. \tag{3.14}$$

Specifically, each frame contains $N_f = N + N_c$ symbols, where N and N_c are the length of data and zero-padding (ZP) sequences. Although the introduction of ZP ensures inter-frame interference (IFI)-free [11], N_c channel taps remain unresolvable after convoluting with the training sequence. In consequence, channel estimation requires joint processing across all taps, leading to high storage demand and heavy computational burden. More importantly, exploiting the delay-domain sparsity becomes an intractable task.

To avoid these limitations, devising a new training pattern would be critical. Specifically, this new pattern should not only facilitate channel estimation but also keep friendly to implementation. To this end, a new training pattern is designed as follows [18]

$$[s(0), s(1), \cdots, s(N_f-1)] =$$

$$\Big[\underbrace{s_0, \overbrace{0, \cdots, 0}^{N_c-1}}_{(1)} \Big| \underbrace{s_1, \overbrace{0, \cdots, 0}^{N_c-1}}_{(2)} \Big| \cdots\cdots \Big| \underbrace{s_{L-1}, \overbrace{0, \cdots, 0}^{N_c-1}}_{(L)}\Big]. \tag{3.15}$$

To be specific, each frame is divided into $L = N_f/N_c$ subframes.[1] Owing to the ZP sequence inlaid with each subframe, sufficient buffer time is left for reconfigurating the APS network after transmitting a non-zero training symbol. In other words, the probing vectors can be updated L times per frame, i.e.,

$$\boldsymbol{p}_{t/r}(n) = \boldsymbol{p}_{t/r}\big(N_c \lfloor n/N_c \rfloor\big), \forall n \in [0, N_f). \tag{3.16}$$

An interesting fact is that, when it comes to the frequency-selective channel estimation in conventional MIMO setup, the training pattern in Eq. (3.15) has been proved optimal in the sense of both the mean squared error (MSE) and system mutual information [19]. Before taking a closer look at the I-O relationship with the new pattern, we first make the following definition.

Random-Probing Vector: At the random-probing stage, the probing vectors are generated by randomly adjusting the angle of each APS component from $\mathcal{B}$. The resultant vector is termed as the random-probing vector and denoted as

[1] Without loss of generality, L is assumed to be an integer here. If N_f/N_c is not an integer, one can simply use $L = \lfloor N_f/N_c \rfloor$.

$$\boldsymbol{p}_{t/r}^{R}(l) = \boldsymbol{p}_{t/r}(lN_c + n_c), (l < L, n_c < N_c). \tag{3.17}$$

Applying random probing is simply because no prior CSI is available at this stage. Note that, the above definition implies that $\boldsymbol{p}_{t/r}^{R}$ possesses both the randomness and subframe-adaptability property. Applying a similar notational change to $\overline{\boldsymbol{p}}_{t/r}$, $\overline{\boldsymbol{P}}_r^*$ becomes

$$\overline{\boldsymbol{P}}_r^* = \begin{bmatrix} \boldsymbol{I}_{N_c} \otimes \left(\overline{\boldsymbol{p}}_r^R(0)\right)^* & \boldsymbol{0} & \cdots & \boldsymbol{0} \\ \boldsymbol{0} & \boldsymbol{I}_{N_c} \otimes \left(\overline{\boldsymbol{p}}_r^R(1)\right)^* & \cdots & \boldsymbol{0} \\ \vdots & \vdots & \ddots & \boldsymbol{0} \\ \boldsymbol{0} & \boldsymbol{0} & \cdots & \boldsymbol{I}_{N_c} \otimes \left(\overline{\boldsymbol{p}}_r^R(L-1)\right)^* \end{bmatrix}, \tag{3.18}$$

and $\overline{\boldsymbol{P}}_t$ is obtained likewise. Substituting the new $\overline{\boldsymbol{P}}_r^*$ and $\overline{\boldsymbol{P}}_t$, together with the training frame into Eq. (3.13), the received signal becomes

$$y(lN_c + n_c) = \left(\overline{\boldsymbol{p}}_r^R(l)\right)^* \overline{\boldsymbol{H}}_{n_c}(lN_c + n_c)\overline{\boldsymbol{p}}_t^R(l)s_l + \xi(lN_c + n_c). \tag{3.19}$$

Clearly, the received samples are now associated with a single channel tap. Hence, our proposed pattern facilitates separating channel taps and thus rendering it possible to exploit the delay-domain sparsity easily. Note that, the success of tap separation does not rely on the non-zero training symbol s_l, therefore we set its quantity as 1 without loss of generality.

3.3.2 *Identification of Effective Taps*

We gather all $\overline{\boldsymbol{H}}_d$-related samples, i.e., $y(lN_c+d)$, $\forall l \in [0, L-1]$, to determine the tap's existence. If at least one dominant path exists therein, $y(lN_c + d)$ will include both the signal and noise parts. Otherwise, it is pure noise. Hence, the detection problem is a binary hypothesis testing problem that can be dealt with via a naive energy detector. By averaging the power of all samples associated with $\overline{\boldsymbol{H}}_d$, we get the test statistics (TS) and its normalized version nTS as follows:

$$Y_d = \frac{1}{L}\sum_{l=0}^{L-1} \left|y(lN_c + d)\right|^2 \tag{3.20a}$$

$$\overline{Y}_d = \frac{Y_d - \sigma^2}{\max\limits_{0 \le m < N_c} (Y_m - \sigma^2, 0)}. \tag{3.20b}$$

When applying CS, random probing is necessary in estimating both the time-invariant and time-varying channels. While for the latter, another important function of random probing is to remain robust against Doppler.

Proposition 1 (Validity of Test Statistics with Doppler) *With sufficient random probings, the test statistics Y_d is approximately irrelevant to the channel's time variation.*

The proof can be made as follows: Let $n = lN_c + d$ and $g_{d,p}(n)$ be the p-th element of $\mathbf{g}_d(n)$, then

$$
\begin{aligned}
y(n) &= \sum_{d=0}^{N_c-1} \overline{\boldsymbol{p}}_r^*(n)\overline{\boldsymbol{H}}_d(n)\overline{\boldsymbol{p}}_t(n-d) + \xi(n) \\
&= \sum_{p=1}^{P} \left(\boldsymbol{p}_r^R(l)\right)^* \boldsymbol{a}_r(\theta_p)\boldsymbol{a}_t^*(\phi_p)\boldsymbol{p}_t^R(l) g_{d,p}(n) + \xi(n).
\end{aligned} \tag{3.21}
$$

Denote $\rho_p(l) = (\boldsymbol{p}_R^R(l))^* \boldsymbol{a}_r(\theta_p)\boldsymbol{a}_t^H(\phi_p)\boldsymbol{p}_T^R(l)$. By using $g_{d,p}(n) = g_{d,p}(0)e^{j\omega_p n}$, we have

$$
\begin{aligned}
|y(n)|^2 = &\sum_{p=1}^{P} |\rho_p(n) g_{d,p}(0)|^2 + 2\mathcal{R}\left\{\sum_{p=1}^{P} \rho_p(l) g_{d,p}(0)\xi(n)e^{j\omega_{p_1}n}\right\} \\
&+ 2\mathcal{R}\left\{\sum_{p_1}\sum_{p_2} \rho_{p_1}(l) g_{d,p_1}(0)\rho_{p_2}^*(l) g_{d,p_2}^*(0)e^{j(\omega_{p_1}-\omega_{p_2})n}\right\}.
\end{aligned} \tag{3.22}
$$

Since $\boldsymbol{p}_t^R$ and $\boldsymbol{p}_r^R$ are random probing vectors with zero mean, it can be readily verified $\mathbb{E}\{\rho_p(l)\} = 0, \forall p \in [1, P]$. By averaging sufficient $|y(n)|^2$ terms, the last two terms in Eq. (3.22) approach zero, thus the TS becomes irrelevant to ω_p.

Proposition 1 ensures the rationale of exploiting the delay-domain sparsity regardless of Doppler effects. Then, energy detection will leave few effective taps as

$$
\mathcal{P}_1 = \left\{d \,\middle|\, \overline{Y}_d \geq \mu\right\} \bigcap \left\{d \,\middle|\, Y_d > \sigma^2\right\} \tag{3.23}
$$

where μ is the threshold.[2] To avoid extreme cases ($\mathbf{cal}(\mathcal{P}_1)$ is either 0 or unreasonably large), a tuning procedure is added, leading to an ultimately selected set as

[2] Evidently, the energy detector is somewhat heuristic. Recall that the energy detector actually plays the role of a binary classifier, a promising direction is to seek the power of deep neural networks. Specifically, given the channel model, a bunch of synthesized data can be generated to train the network for classification (tap detection) in a supervised manner. The offline trained network could then be used for online prediction.

$$\mathcal{P}=\begin{cases}\mathcal{P}_1, & 0<\mathbf{cal}(\mathcal{P}_1)\leq A\\ \left\{d|\overline{Y}_d\geq\overline{\lambda}_A\right\}, & \mathbf{cal}(\mathcal{P}_1)>A\\ \left\{d|Y_d\geq\lambda_A\right\}, & \mathbf{cal}(\mathcal{P}_1)=0.\end{cases} \tag{3.24}$$

with λ_A and $\overline{\lambda}_A$ representing the A-th largest TS and nTS, respectively.

3.3.3 Identification of Effective Beams

Suppose n_c out of N_c taps are recognized effective after tap detection, with their indices denoted by $\mathcal{D}=\{d_1, d_2, \cdots, d_D\}$. We then proceed to determine the angle support for those taps belonging to $\mathcal{D}$. The samples we use are already obtained at the random-probing stage, so this step does not require extra training frames.

Take tap-d_i ($d_i \in \mathcal{D}$) for example and omit the subscript of d_i for brevity. Let us first derive the sparse representation for received samples in order to apply CS. Stacking all $\overline{\boldsymbol{H}}_d$-related samples from $\boldsymbol{y}$ yields

$$\boldsymbol{y}_d=[y(d), y(N_c+d), \cdots, y((L-1)N_c+d)]'. \tag{3.25}$$

Denoting $n_l = lN_c + d$ $\left(\forall l \in [0, L)\right)$ and using matrix equality $\text{vec}(\boldsymbol{ABC})=(\boldsymbol{C}'\otimes\boldsymbol{A})\text{vec}(\boldsymbol{B})$, $y(n_l)$ can be rewritten as

$$y(n_l)=\underbrace{\left((\overline{\boldsymbol{p}}_t^R(l))'\otimes(\overline{\boldsymbol{p}}_r^R(l))^*\right)}_{\boldsymbol{\psi}(l)}\underbrace{\left(\mathbf{vec}(\overline{\boldsymbol{H}}_d(n_l))\right)}_{\overline{\boldsymbol{h}}_d(n_l)}+\xi(n_l). \tag{3.26}$$

By neglecting the noise part, we can express $\boldsymbol{y}_d$ as

$$\boldsymbol{y}_d=\underbrace{\begin{bmatrix}\boldsymbol{\psi}(0) & \mathbf{0} & \cdots & \mathbf{0}\\ \mathbf{0} & \boldsymbol{\psi}(1) & \cdots & \mathbf{0}\\ \vdots & \vdots & \ddots & \vdots\\ \mathbf{0} & \mathbf{0} & \cdots & \boldsymbol{\psi}(L-1)\end{bmatrix}}_{\boldsymbol{\Psi}}\underbrace{\begin{bmatrix}\overline{\boldsymbol{h}}_d(n_0)\\ \overline{\boldsymbol{h}}_d(n_1)\\ \vdots\\ \overline{\boldsymbol{h}}_d(n_{L-1})\end{bmatrix}}_{\overline{\boldsymbol{h}}_d}. \tag{3.27}$$

In a special case of time-invariant channels, all $\overline{\boldsymbol{h}}_d(n_l)$'s turn out to be exactly equal, giving rise to

$$\mathbf{y}_d = \begin{bmatrix} \boldsymbol{\psi}(0) \\ \boldsymbol{\psi}(1) \\ \vdots \\ \boldsymbol{\psi}(L-1) \end{bmatrix} \overline{\mathbf{h}}_d(n_0). \tag{3.28}$$

Determining the angle support of $\overline{\mathbf{H}}_d$ is equivalent to locating those non-zero entries from the $G_t G_r$-dimensional vector $\overline{\mathbf{h}}_d$. This can be readily done via OMP, through which $\mathcal{O}(P \log G_t G_r)$ rather than $\mathcal{O}(G_t G_r)$ samples are needed to guarantee accuracy.

Unfortunately, Eq. (3.28) becomes invalid in the presence of Doppler shifts, motivating us to restudy Eq. (3.27). Since $\overline{\mathbf{h}}_d$ remains sparse for $LP \ll LG_t G_r$, a natural option would still be OMP. Reminisce that the angle variations are negligible during channel estimation, indicating an angle support is shared by all $\overline{\mathbf{h}}_d(n_l)$'s. But this unique structure cannot be exploited by OMP because $\mathbf{h}_d(n_l)$ will be treated as a generic sparse vector in implementation. Considering this, a more general block-sparse representation will be derived to utilize this unique property. Specifically, we first construct a permutation matrix $\mathbf{P}$, satisfying $\mathbf{P}[:, (i-1)G_t G_r + j] = \mathbf{I}_{G_t G_r}[:, (j-1)G_t G_r + i]$ [9]. Then $\mathbf{y}_d$ can be decomposed as

$$\mathbf{y}_d = \left(\boldsymbol{\Psi} \mathbf{P}\right) \cdot \left(\mathbf{P}' \overline{\mathbf{h}}_d\right). \tag{3.29}$$

The "new" sparse signal and sensing matrix accordingly become

$$\widetilde{\mathbf{h}}_d = \mathbf{P}' \overline{\mathbf{h}}_d = \left[\widetilde{\mathbf{h}}_{d,1}', \widetilde{\mathbf{h}}_{d,2}', \cdots, \widetilde{\mathbf{h}}_{d,G_t G_r}'\right]' \tag{3.30a}$$

$$\widetilde{\boldsymbol{\Psi}} = \boldsymbol{\Psi} \mathbf{P} = \left[\widetilde{\boldsymbol{\Psi}}_1, \widetilde{\boldsymbol{\Psi}}_2, \cdots, \widetilde{\boldsymbol{\Psi}}_{G_t G_r}\right] \tag{3.30b}$$

where $\widetilde{\mathbf{h}}_{d,i} = \left[\overline{h}_{d,i}(n_0), \cdots, \overline{h}_{d,i}(n_{L-1})\right]^*$ and $\widetilde{\boldsymbol{\Psi}}_i = \mathbf{diag}\left[\psi_i(0), \cdots, \psi_i(L-1)\right]$, $\forall i \in [1, G_t G_r]$, with $\overline{h}_{d,i}(n_j)$ and $\psi_i(l)$ being the i-th entry of $\overline{\mathbf{h}}_d(n_j)$ and $\boldsymbol{\psi}(l)$, respectively.

Different from the original $\overline{\mathbf{h}}_d$, the rearranged $\widetilde{\mathbf{h}}_d$ exhibits block sparsity [20], and the sparsity level equals to that of the $\overline{\mathbf{h}}_d(n_l)$.

Block OMP appears to be an efficient tool to estimate $\widetilde{\mathbf{h}}_d$ in Eq. (3.27). Nevertheless, reaching this goal still encounters two major difficulties:

P.1 How to properly set iterations for CS algorithms.
P.2 How to combat strong Doppler effects.

To address these problems, we propose an algorithm termed as adaptive-block OMP (A-BOMP) algorithm. Specifically, when applying CS, iterations should be set as equal to (or a slightly higher than) the signal sparsity. Considering the actual sparsity is generally unknown, most works will adopt a large value to reduce the risks of estimation loss. However, once the iterations significantly mismatch the signal sparsity, computational complexity and potential over-fitting errors may surge. Albeit not knowing the sparsity either, we will show that, a proper iteration setting is still possible,

Recall that D out of N_c taps have been regarded effective, indicating at most D beams are there. Thus, the signal sparsity is upper bounded by D. Actually, such an upper bound can be further lowered by leveraging the following result.

Lemma 1 *For the wideband channel with N_c taps, the probability that k out of K ($k \leq K$) beams reside within one tap is approximated as*

$$P(K,k) = C_K^k \left(\frac{1}{N_c}\right)^k \left(\frac{N_c - 1}{N_c}\right)^{K-k}. \tag{3.31}$$

A brief illustration is made for $N_c = 128$ and $K = 10$. In this case, $P(10, 4) < 10^{-5}$, implying that it is virtually impossible for one tap to contain more than 3 beams, regardless to say 10 for $P(10, 10) < 10^{-18}$. Combing all these discussions, a proposition is officially made below to address P.1.

Proposition 2 (Number of Iterations) *Let P_T be a small threshold (e.g., 10^{-3}) and D be the number of effective taps after tap identification. A proper iterations can be set as $k - 1$, where k is the smallest integer satisfying $P(D, k) < P_T$.*

To tackle P.2, how to utilize the block structure is critical [21]. One solution was proposed in [22] by using the BEM model to capture the variations. Despite dramatically lowering deterioration, this approach exhibits two drawbacks. First, its performance will be heavily dependent on the basis order. Secondly, it involves a large-scale eigenvalue decomposition (EVD) with complexity $\mathcal{O}(L^3)$). Luckily, the proposed A-BOMP, whose pseudo-code is presented in **Algorithm 1**, can bypass these issue. When implementing the algorithm, each outer iteration consists of three parts:

S$_1$ *partial basis matching*: select an angle pair having the largest sum of grouping correlations

S$_2$ *resolution refinement*: re-construct sensing matrix associated with the selected angle pair and implement refine resolution.

S$_3$ *partial residue update*: estimate the coefficients by the least-squared (LS) estimator and then apply residue update.

The group size S (equivalent to the group number G) is a key parameter to A-BOMP. In the presence of Doppler, the non-zero support size always exceeds the number of measurements, making localizing the non-zero support a great challenge, not to mention recovering the entire vector. The severe shortage of measurements forces us to "shrink" the non-zero support. To this end, we divide $\widetilde{\boldsymbol{h}}_{d,i}$ defined in Eq. (3.30a) into S groups. For those entries belonging to one group, they are highly correlated thus being treated equally. In consequence, the group division essentially performs signal compression.

Algorithm 1 Proposed A-BOMP algorithm

Require: *Received signal* $\mathbf{y}_d$ *and sensing matrix* $\widetilde{\boldsymbol{\Psi}}_d$, *maximum block-sparsity* $\mathcal{K}$, *group size* S, *group number* $G = \frac{L}{S}$, *and error threshold* ϵ.

Ensure: *The AoA support set* $\widetilde{\mathcal{A}}_d$ *and the corresponding AoD support set* $\widetilde{\mathcal{D}}_d$.

1: *Initialization: The residue* $\mathbf{r}_d = \mathbf{y}_d$, *iteration index* $C = 0$, $\mathcal{A}_d/\widetilde{\mathcal{A}}_d$ *and* $\mathcal{D}_d/\widetilde{\mathcal{D}}_d$ *are set to be empty,* $\beta = \infty$, $\boldsymbol{\Phi} = \varnothing$ *and* $x = x_0 = 0$.
2: **while** $C < \mathcal{K}$ and $\beta > \epsilon$ **do**
3: $\quad C = C + 1;$
4: $\quad g_i = \arg\max_{g} \sum_{i=1}^{G} \frac{\|[\widetilde{\boldsymbol{\Psi}}^*_{d,g}\mathbf{r}_d]((i-1)S+1:iS)\|_1}{\|\widetilde{\boldsymbol{\Psi}}_{d,g}\|_F}$
5: $\quad n_R = \lceil g_i/G_t \rceil$ and $n_T = g_i - (n_R - 1)G_t$.
6: $\quad$ **if** $\exists\, i,\ \bmod\,(|n_D/n_R - \mathcal{D}(i)/\mathcal{A}(i)|, G_t/G_r) \leq 1$ **then**
7: $\qquad$ **goto** 2
8: $\quad$ **end if**
9: $\quad \mathcal{A} = \{\mathcal{A}, n_R\}, \mathcal{D} = \{\mathcal{D}, n_T\}$
10: $\quad \widehat{\mathbf{A}}_T = [\boldsymbol{f}_{N_t}(\frac{n_T-1}{G_t} + \frac{2j_T}{G_t^2})]_{j_T \in [-\frac{G_t}{2}:1:\frac{G_t}{2}-1]}$
11: $\quad \widehat{\mathbf{A}}_R = [\boldsymbol{f}_{N}(\frac{n_R-1}{G_r} + \frac{2j_R}{G_R^2})]_{j_R \in [-\frac{G_r}{2}:1:\frac{G_r}{2}-1]}$
12: $\quad \widehat{\boldsymbol{\psi}}[n_i] = (\boldsymbol{p}'_A[n_i] \otimes \boldsymbol{w}^*_A[n_i])(\widehat{\mathbf{A}}^*_T \otimes \widehat{\mathbf{A}}_R)_{i=1\sim L}$
13: $\quad \widehat{\boldsymbol{\Psi}}_{d,i} = \mathbf{diag}\big[\widehat{\boldsymbol{\psi}}[n_1](i), \cdots, \widehat{\boldsymbol{\psi}}[n_{G_tG_r}](i)\big]_{i=1\sim G_tG_r}$
14: $\quad \widehat{g}_i = \arg\max_{g} \sum_{i=1}^{G} \frac{\|[\widehat{\boldsymbol{\Psi}}^*_{d,g}\mathbf{r}_d]((i-1)S+1:mS)\|_1}{\|\widehat{\boldsymbol{\Psi}}_{d,g}\|_F}$
15: $\quad \widehat{n}_R = \lceil \widehat{g}_i/G_t \rceil$ and $\widehat{n}_T = \widehat{g}_i - (\widehat{n}_R - 1)G_t$.
16: $\quad \boldsymbol{\Phi} = \big[\boldsymbol{\Phi}, \boldsymbol{M}(\boldsymbol{f}^*_{N_t}(\frac{G_t(n_T-1)+\widehat{n}_T}{G_t^2}) \otimes \boldsymbol{f}_{N_r}(\frac{G_r(n_R-1)+2\widehat{n}_R}{G_r^2})\big]$
17: $\quad$ **for** $j = 1 : G$ **do**
18: $\qquad \boldsymbol{j} = (j-1)S + 1 : mS$
19: $\qquad x = x + \|\boldsymbol{\Phi}(\boldsymbol{j},:)^{\dagger}\boldsymbol{y}(\boldsymbol{j})\|_2$
20: $\qquad \boldsymbol{r}_d(\boldsymbol{j}) = \boldsymbol{y}_d(\boldsymbol{j}) - \boldsymbol{\Phi}(\boldsymbol{j},:)\boldsymbol{\Phi}(\boldsymbol{j},:)^{\dagger}\boldsymbol{y}(\boldsymbol{j})$
21: $\quad$ **end for**
22: $\quad \beta = |x - x_0|/x$
23: $\quad x_0 = x, x = 0$
24: $\quad \widetilde{\mathcal{A}}_d = \big\{\widetilde{\mathcal{A}}_d, \frac{2\pi G_r(n_R-1)+\widehat{n}_R}{G_r^2}\big\}$
25: $\quad \widetilde{\mathcal{D}}_d = \big\{\widetilde{\mathcal{D}}_d, \frac{2\pi G_t(n_T-1)+\widehat{n}_T}{G_t^2}\big\}$
26: **end while**

Proposition 3 (Determination of the Group Size) *Let* τ *denote a high-correlation value (e.g., 0.707). A proper group size can be set as the largest* S *meeting* $\cos(\omega_{max} N_c S) \leq \tau$ *and* $\omega_{max} N_c S \leq \pi/2$.

Proposition 3 indicates that a smaller ω_{max} results in a larger S. For $\omega_{max} = 0$, i.e., a time-invariant channel, A-BOMP degenerates into BOMP. Compared to the latter, A-BOMP only introduces a few small-scale matrix inversions. Simulations show that such minimal computational cost will bring in a significantly improved accuracy.

After running A-BOMP, the steering matrices for tap-d channel are estimated as

$$\widetilde{\mathbf{A}}_{r,d} = \big[\boldsymbol{f}_{N_r}\big(\widetilde{\mathcal{A}}_d(1)/2\pi\big), \cdots, \boldsymbol{f}_{N_r}\big(\widetilde{\mathcal{A}}_d(c_d)/2\pi\big)\big] \tag{3.32a}$$

$$\widetilde{\boldsymbol{A}}_{t,d}=\left[\boldsymbol{f}_{N_t}\left(\widetilde{\mathcal{D}}_d(1)/2\pi\right),\cdots,\boldsymbol{f}_{N_t}\left(\widetilde{\mathcal{D}}_d(c_d)/2\pi\right)\right] \tag{3.32b}$$

with $c_d = \mathbf{cal}(\widetilde{\mathcal{A}}_d)$. The approximate beamspace representation for tap-d channel bear a form as

$$\widetilde{\boldsymbol{H}}_d(n) = \widetilde{\boldsymbol{A}}_{r,d}\mathbf{diag}(\widetilde{\boldsymbol{g}}_d(n))\widetilde{\boldsymbol{A}}^*_{t,d}. \tag{3.33}$$

where $\widetilde{\boldsymbol{g}}_d(n)$ consists of the unknown path gains.

3.3.4 *Identification of Beam Amplitudes*

In the final stage, steered-probing will be implemented based on the estimated beam direction to accurately estimate the path gains and Doppler shifts. Specifically, for tap-d channel, let us construct a set $\mathcal{I}_d$ whose i-th element is $(\mathcal{A}_d(i), \mathcal{D}_d(i))$. Due to off-grid leakage, different $\mathcal{I}_d$'s may share the same element, so we union them as

$$\mathcal{I} = \mathcal{I}_{d_1}\bigcup\mathcal{I}_{d_2}\bigcup\cdots\bigcup\mathcal{I}_{d_D}. \tag{3.34}$$

All AoAs and AoDs are individually extracted from $\mathcal{I}$ and captured by $\mathcal{I}_A$ and $\mathcal{I}_D$. To facilitate beamforming, only a bunch of discrete AoD indices will be reported to the Tx-end. Without triggering ambiguity, we reset the time instant at this stage. Some definitions are made in order.

Steered-Probing Vector *We define* $\boldsymbol{p}_t^S(n)$ *and* $\boldsymbol{p}_t^S(n)$ *to be the RF vectors at time instant* n. *To improve receive SNR,* $p^S_{p,t}(n)$ *(the* p*-th element of* $\boldsymbol{p}_t^S(n)$*) and* $p^S_{q,r}(n)$ *(the* q*-th element of* $\boldsymbol{p}_r^S(n)$*) are designed as*

$$p^S_{p,t}(n) = \frac{1}{\sqrt{N_t}}e^{jQ\left((p-1)\mathcal{I}_D(\widehat{n})\right)}, p \in [1, N_t] \tag{3.35a}$$

$$p^S_{q,r}(n) = \frac{1}{\sqrt{N_r}}e^{jQ\left((q-1)\mathcal{I}_A(\widehat{n})\right)}, q \in [1, N_r] \tag{3.35b}$$

with $\widehat{n} = \bmod\left(\lfloor n/N_c\rfloor, cal(\mathcal{I})\right)$.

At the i-th polling, we stack all the samples relating to tap-d:

$$\boldsymbol{y}_{d,i} = \left[y\left(d+n_{i,0}\right),\cdots,y\left(d+n_{i,|\mathcal{I}|-1}\right)\right]' \tag{3.36}$$

where $n_{i,j} = N_c j + \mathbf{cal}(\mathcal{I})N_c i, \forall j \in [0, cal(\mathcal{I}))$. Following Eq. (3.33), each sample in $\boldsymbol{y}_{d,i}$ can be approximated as

$$\begin{aligned} y(n_{i,j}+d) &\approx (\boldsymbol{p}_r^S(n_{i,j}))^*\widetilde{\boldsymbol{A}}_{r,d}\mathbf{diag}(\widetilde{\boldsymbol{g}}_d(n_{i.j}+d)) \times \widetilde{\boldsymbol{A}}^*_{t,d}\boldsymbol{p}_t^S(n_{i,j}) + \xi(n_{i,j}+d) \\ &= \mathbf{vec}'\left(\mathbf{diag}(\widetilde{\boldsymbol{g}}_d(n_{i,j}+d))\right)\boldsymbol{m}_d(n_{i,j}) + \xi(n_{i,j}+d) \end{aligned} \tag{3.37}$$

where $\boldsymbol{m}_d(n_{i,j}) = \left((\boldsymbol{p}_t^S(n_{i,j}))'\widetilde{\boldsymbol{A}}_{t,d}^*\right) \otimes \left((\boldsymbol{p}_r^S(n_{i,j}))^*\widetilde{\boldsymbol{A}}_{r,d}\right)$. By capturing the amplitudes corresponding to the sample in the middle, $\boldsymbol{y}_{d,i}$ can be approximately represented as

$$\boldsymbol{y}_{d,i} \approx \underbrace{\begin{bmatrix} \boldsymbol{m}_d(n_{i,0}) \\ \boldsymbol{m}_d(n_{i,1}) \\ \vdots \\ \boldsymbol{m}_d(n_{i,cal(\mathcal{I})-1}) \end{bmatrix}}_{\boldsymbol{M}_{d,i}} \text{vec}\Big(\textbf{diag}\big(\boldsymbol{g}_d(\overline{n}_i)\big)\Big) + \boldsymbol{\xi}_{d,i} \tag{3.38}$$

with $\overline{n}_i = \left(n_{i,0} + n_{i,\textbf{cal}(\mathcal{I})-1}\right)/2 + d$ and

$$\boldsymbol{\xi}_{d,i} = [\xi(n_{i,0}+d), \xi(n_{i,1}+d), \cdots, \xi(n_{i,\textbf{cal}(\mathcal{I})-1}+d)]'.$$

Let $\widetilde{\boldsymbol{M}}_{d,i} = \left[\boldsymbol{M}_{d,i}[:, 1^2], \boldsymbol{M}_{d,i}[:, 2^2]\cdots, \boldsymbol{M}_{d,i}[:, C_d^2]\right]$, then Eq. (3.38) equals to

$$\boldsymbol{y}_{d,i} \approx \widetilde{\boldsymbol{M}}_{d,i}\boldsymbol{g}_d(\overline{n}_i) + \boldsymbol{\xi}_{d,i} \tag{3.39}$$

Since $\textbf{cal}(\mathcal{I}) \geq c_d$, $\boldsymbol{g}_d(\overline{n}_i)$ can be recovered by the (least-squares) LS estimator:

$$\widehat{\boldsymbol{g}}_d(\overline{n}_i) = \widetilde{\boldsymbol{M}}_{d,i}^{\dagger}\boldsymbol{y}_{d,i} = \boldsymbol{g}_d(\overline{n}_i) + \widetilde{\boldsymbol{M}}^{\dagger}\boldsymbol{\xi}_{d,i}. \tag{3.40}$$

Once getting a new $\widehat{\boldsymbol{g}}_d$, we pick its j-th element that corresponds to the beam-j's estimated amplitudes in current polling. After $R = \lfloor L/\textbf{cal}(\mathcal{I})\rfloor$ rounds of polling,[3] we get

$$\widehat{\boldsymbol{g}}_{d,j} = \left[\widehat{g}_{d,j}(\overline{n}_0), \widehat{g}_{d,j}(\overline{n}_1), \cdots, \widehat{g}_{d,j}(\overline{n}_{R-1})\right]'. \tag{3.41}$$

Lemma 2 *Through repetitive polling, the pseudo time series $\widehat{\boldsymbol{g}}_{d,j}$ has an equal sampling interval thus can be modeled as finite noisy samples of a single-tone sinusoid.*

Many techniques have been proposed over the years for estimating the frequency of a complex sinusoid in additive white Gaussian noise. Here we adopt the WNALP estimator that is known for computational efficiency and near optimality [23]. The detailed procedures are described as below

- *Set $M_0 = \lfloor R/2 \rfloor$.*

[3] Similar to the random-probing stage, we introduce the steered-probing state based on one frame consisting of L subframes. In practice or numerical comparisons, one can simply replace L with the actual number of subframes, i.e., $\textbf{cal}(I)R$.

- *Calculate the auto-correlation of* $\widehat{\mathbf{g}}_{d,j}$ *as*

$$R(m) = \frac{1}{R-m} \sum_{i=m+1}^{2M_0} \widehat{g}_{d,j}(\overline{n}_i)\widehat{g}^*_{d,j}(\overline{n}_{i-m}) \tag{3.42}$$

- *Calculate the smoothing coefficient* w_m *as*

$$w_m = \frac{3\big((M_0 - m)(2M_0 - m + 1) - M_0^2\big)}{M_0(4M_0^2 - 1)} \tag{3.43}$$

- *Estimate the Doppler shift as*

$$\widehat{\omega}_{d,j} = \frac{1}{N_c cal(\mathcal{I})} \sum_{m=1}^{M_0} w_m \mathbf{angle}(R(m)R^*(m-1)) \tag{3.44}$$

- *Estimate the amplitudes as*

$$\begin{aligned}\widehat{g}_{d,j}(d) &= \frac{e^{-j\widehat{\omega}_{d,j}\frac{N_c cal(\mathcal{I})}{2}}}{R} \sum_{i=1}^{R} \widehat{g}_{d,j}(\overline{n}_i) e^{-j\widehat{\omega}_{d,j} N_c cal(\mathcal{I})(i-1)} \\ &= \frac{1}{R} \sum_{i=1}^{R} \widehat{g}_{d,j}(\overline{n}_i) e^{-j\widehat{\omega}_{d,j} N_c cal(\mathcal{I})(i-1/2)}.\end{aligned} \tag{3.45}$$

The rest beams can be estimated similarly thus being omitted.

3.4 Simulations

In this section, extensive numerical results are presented to verify the advantages of the proposed approach over existing works. In simulations, the system carrier frequency f_c is 60 GHz. The number of antennas is $N_t = N_r = 32$, The dictionary sizes is $G_t = G_r = 64$. $h(\cdot)$ is the raised-cosine filter with the roll-off factor $\beta = 1$. Each channel realization is generated according to Eq. (3.3) with P ranging from 1 to 4. One-stage refinement is applied for all cases when applying OMP-based methods. If not specified, the resolution of APS is 2-bit. Other simulation parameters include $N_c = 16$, $N = 64$, $T_s = 50ns$, $A = 8$, $P_T = 10^{-3}$, $\epsilon = 0.01$ and $\mu = 0.03$. The SNR (averaged TSNR) is defined as $\frac{L}{N\sigma^2}$.

The estimation performance is evaluated based on the normalized MSE (NMSE) which is defined as

$$\varepsilon = \frac{\sum_{d=0}^{N_c-1} \parallel \boldsymbol{H}_d - \widehat{\boldsymbol{H}}_d \parallel_F^2}{\sum_{d=0}^{N_c-1} \parallel \boldsymbol{H}_d \parallel_F^2}. \tag{3.46}$$

Each curve is on the average of 1000 channel realizations.

3.4.1 Tap Identification

To verify the effectiveness of tap identification, we plot the averaged selected taps as well as their power ratio in Fig. 3.2. We set $P = 3$ and allocate 40 frames at the random-probing stage. Three different v_m's: 0, 12 and 120 km/h are considered. We see that the tap identification is regardless of Doppler effects. As SNR increases, fewer taps will be selected, and this reduction in taps can be up to 75% at 0dB. As the pulse-shaping filter uses a raised-cosine function, the delay-domain also suffers from off-grid issues due to side-lobe leakage, making the identified taps are slightly over actual paths. Note that, the large reduction in taps to be processed is not at the cost of power loss. As can be seen from Fig. 3.2, the averaged power ratio soon exceeds 97% at medium SNR. The effectiveness of tap identification is attributed to the delay-domain sparsity of mmWave channels.

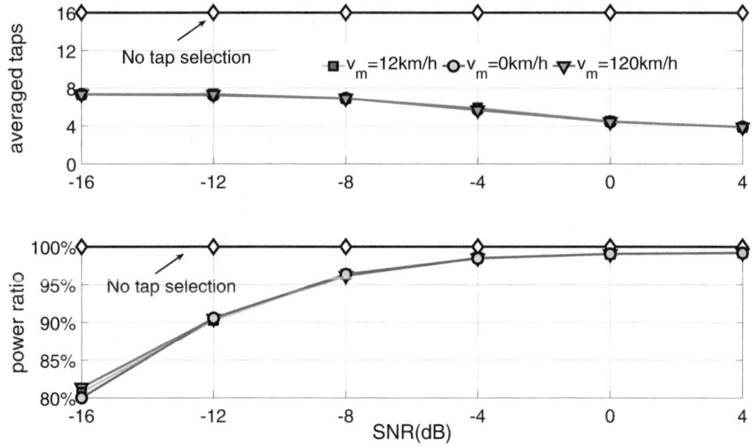

Fig. 3.2 The averaged power ratio contained by the selected taps versus all taps

3.4.2 NMSE in Static Wideband Channels

In this part, we start by comparing the doubly-sparse approach (DSA) with state-of-the-art beamspace-sparse approach (BSA) [11]. The channel is generated with 3 paths and $\omega_m = 0$. For DSA, 40 training frames are allocated at the random-probing stage and 4-time polling is operated at the steered-probing stage. For BSA, 60 training frames are allocate. The sensing matrix in BSA is 16384×131072, requiring a memory space over 18 GB, in contrast to our 200×4096 matrix occupying 9 Mb memory space. Due to the great shortage of training frames, the LS estimator without utilizing any sparsity performs the worst. BSA with regularized LS estimator although performs much better than LS but still much worse than DSA. The insufficient probings make the NMSE curve soon go flat. Even under the same peak SNR, Fig. 3.3 that DSA still outperforms BSA at medium-to-high SNR region. This implies the benefits brought by DSA outweigh the training pattern's power inefficiency.

In Fig. 3.4, we further plot the averaged consumed training frames for different approaches. By observing two figures together, it is clear to see an improper iteration setting ($\mathcal{K} = 8$) will result in additional training overhead without making any substantial performance improvement. Thanks to proposition 2, iterations can be properly set for A-BOMP (A-BOMP is equivalent to OMP here). Therefore the resultant NMSE can be very close to the ideal benchmark ($\mathcal{K} = 4$). With pre-determined iterations, DSA requires the least training overhead, with a reduction of 20% compared to BSA at high SNR.

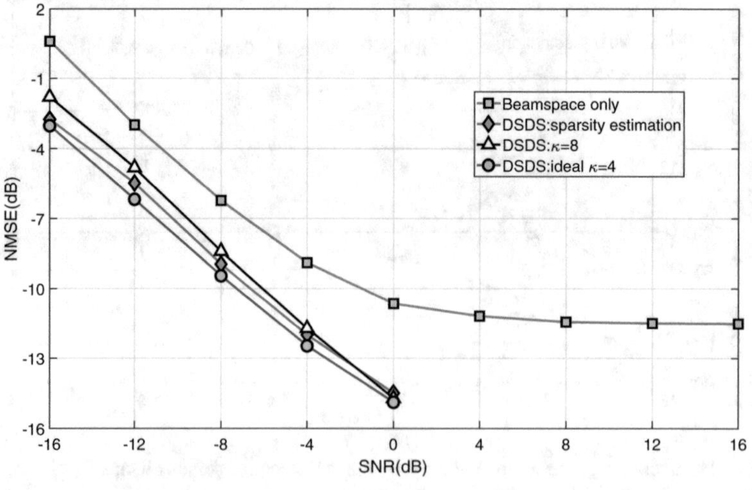

Fig. 3.3 The NMSE comparisons among different schemes in static wideband channels

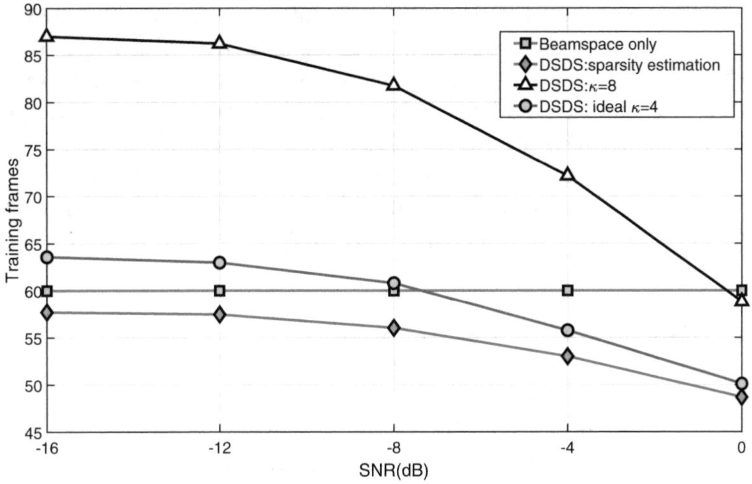

Fig. 3.4 The total training frames consumed by different schemes in static wideband channels

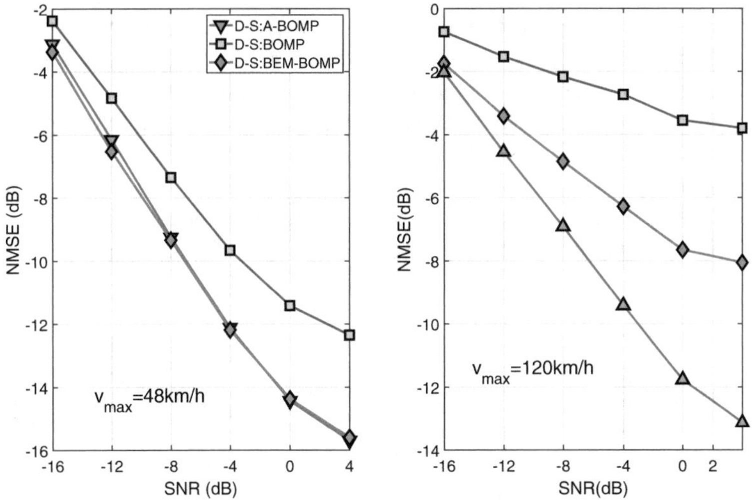

Fig. 3.5 The NMSE comparisons in "frequency-flat" and time-varying channels in modest mobility

3.4.3 *NMSE in Frequency-Flat Time-Varying Channels*

In Fig. 3.5, we consider a frequency-flat but time-varying channel. The channel is generated by setting $P = 3$, and v_m=48 or 120 km/h. We compare the NMSE achieved via the angle support recovered by A-BOMP and DPC-BOMP. As the channel has only one tap, this comparison boils down to comparing with state-of-the-art FTV channel estimator [9]. Specifically, we allocate 60 frames at the random-probing stage and 4-time polling at the steered-probing stage. The order of DPC basis is 2 as in [9]. In modest mobility (v_m=48 km/h), A-BOMP and DPC-BOMP perform similarly, both outperforming BOMP remarkably. In high mobility (v_m=120 km/h), the advantage of A-BOMP over DPC-BOMP is well noticeable because it per-group matching can well undertake the role of unknown compression. Apart from the merit in estimation, A-BOMP is also more computationally efficient because it avoids large-scale EVD in DPC-BOMP.

3.4.3.1 NMSE in Doubly-Selective Channels

To give a thorough evaluation for DSA, we fix SNR=-1 dB and v_m=55 km/h, then simulate NMSE versus frame duration under various configurations.

In Fig. 3.6, we compare NMSE by varying the path number. The remaining parameters are fixed as $p_1 = 60$, $R = 4$, and $b = 2$. The results show that without Doppler compensation NMSE will soon exceed -10 dB, resulting in a great discrepancy with the actual channels. After compensation via the estimates, superb tracking ability can be guaranteed for up to 20 frames. Furthermore, the proposed method show a minimal performance degradation by increasing P from 2 to 4, while a nearly 2 dB degradation is seen by [11] under in a similar setup. This demonstrate that the proposed estimator is more robust against frequency selectivity.

In Fig. 3.7, we compare NMSE by varying the APS's resolution (referring to b). The remaining parameters are fixed as $p_1 = 60$, $b = 2$, $R = 4$, and $P = 3$. The figure illustrates that $b = 1$ leads to a poor performance but increasing it by 1, i.e., using 2-bit APS, will bring a huge improvement. The performance gap compared to

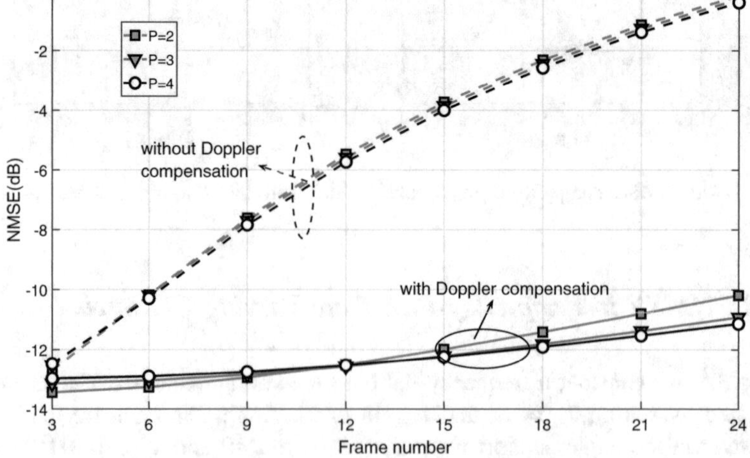

Fig. 3.6 The NMSE versus the number of paths

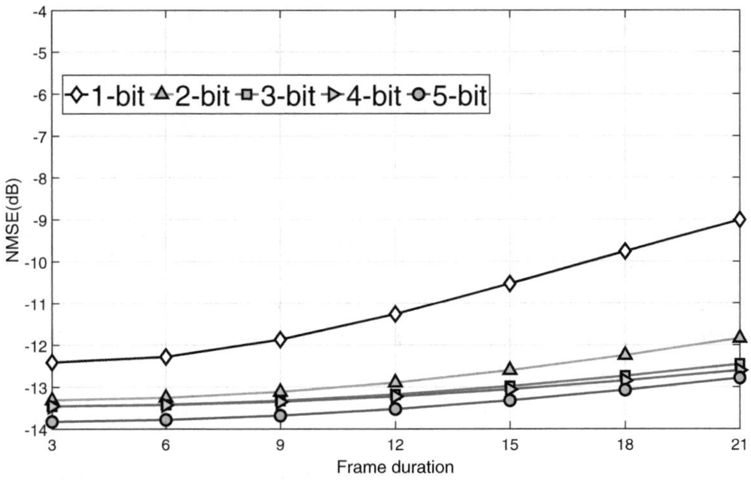

Fig. 3.7 The NMSE versus APS resolution

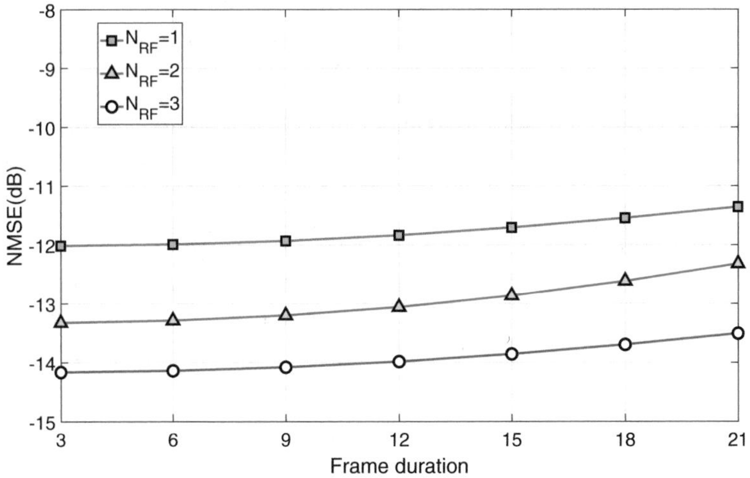

Fig. 3.8 The NMSE versus the number of RF chains

fine-APS cases (3∼5-bit) has been very small (only 0.5 dB), implying that the DSA is suitable to low-bit-APS setup.

In Fig. 3.8, we compare NMSE by varying the number of RF chains. Other parameters are fixed as $P = 3$, $b = 2$, $R = 6$, and $p_1 = 30$. As expected, multiple RF chains can lower the estimation error because they can generate more random probing patterns, which in turn benefit the recovery of angle support. We need to mention that all non-zero symbols are set as one throughout estimation. Actually, if the peak to average power ratio (PAPR) is not a significant concern, one can potentially set these symbols as the Gaussian distributed variables like [24] to reinforce the randomness.

3.5 Discussions and Summary

In this chapter, we investigated the doubly-selective channel estimation for hybrid mmWave mMIMO systems. Compared with existing works, our proposed doubly-sparse approach has been demonstrated to be a more general and superior solution to channel estimation under hybrid mmWave mMIMO. The implementation of the proposed DSDS channel estimator entails four key components:

1. Send multiple impulse-like training pilots to separate channel taps with random probing applied.
2. Identify significant channel taps via energy detection to exploit the delay-domain sparsity regardless of Doppler.
3. Identify the direction of significant beams via A-BOMP with an effective mechanism to combat time-variation.
4. Apply steered-probing to estimate the amplitudes and Doppler using high-quality received samples.

The proposed channel estimator is tailored for general doubly-selective channels. In practice, the investigated channel may not exhibit double selectivity, so one can use part of the above steps to accommodate these special cases. To ensure a reliable recovery via OMP, the sensing matrix should best satisfy the restricted isometric property (RIP). The optimal sensing matrix in terms of the RIP is independently and identically distributed (IID) Gaussian matrix. Unfortunately, due to the constant-modulus limitation of the APS, the optimal sensing matrix design remains an open topic that deserves further investigation.

References

1. F. Boccardi, R.W. Heath, A. Lozano, T.L. Marzetta, P. Popovski, Five disruptive technology directions for 5G. IEEE Commun. Mag. **52**(2), 74–80 (2014)
2. C.-X. Wang, J. Bian, J. Sun, W. Zhang, M. Zhang, A survey of 5G channel measurements and models. IEEE Commun. Surv. Tuts. **20**(4), 3142–3168 (2018)
3. J. Huang, C.-X. Wang, R. Feng, J. Sun, W. Zhang, Y. Yang, Multi-frequency MmWave massive MIMO channel measurements and characterization for 5G wireless communication systems. IEEE J. Sel. Areas Commun. **35**(7), 1591–1605 (2017)
4. A. Alkhateeb, O.E. Ayach, G. Leus, R. Heath, Channel estimation and hybrid precoding for millimeter wave cellular systems. IEEE J. Sel. Topics Signal Process. **8**, 5, 831–846 (2014)
5. C. Chen, Y. Dong, X. Cheng, L. Yang, Low-resolution PSs based hybrid precoding for multi-user communication systems. IEEE Trans. Veh. Technol. **67**(7), 6037–6047 (2018)

6. S. Gao, Y. Dong, C. Chen, Y. Jin, Hierarchical beam selection in mmWave multiuser MIMO systems with one-bit analog phase shifters, in *Proceedings of IEEE International Conference on Wireless Communications and Signal Processing, Yangzhou* (2016)
7. C. Gustafson, K. Haneda, S. Wyne, F. Tufvesson, On mmWave multipath clustering and channel modeling. IEEE Trans. Antennas Propag. **62**(3), 1445–1455 (2014)
8. J.A. Tropp, A.C. Gilbert, Signal recovery from random measure- ments via orthogonal matching pursuit. IEEE Trans. Inf. Theory **53**(12), 4655–4666 (2007)
9. Q. Qin, L. Gui, P. Cheng, Time-varying channel estimation for millimeter wave multiuser MIMO systems. IEEE Trans. Veh. Technol. **67**(10), 9435–9448 (2018)
10. Z. Gao, C. Hu, L. Dai, Z. Wang, Channel estimation for millimeterwave massive MIMO with hybrid precoding over frequency-selective fading channels. IEEE Commun. Lett. **20**(6), 1259–1262 (2016)
11. K. Venugopal, A. Alkhateeb, N. Gonzalez-Prelcic, R. Heath, Channel estimation for hybrid architecture-based wideband millimeter wave systems. IEEE J. Sel. Areas Commun. **35**(9), 1996–2009 (2017)
12. J. Fernandez, N. Prelcic, K. Venugopal, R. Heath, Frequency-domain compressive channel estimation for frequency-selective hybrid mmWave MIMO systems. IEEE Trans. Wireless Commun. **17**(5), 2946–2960 (2018)
13. C. Carbonelli, S. Vedantam, U. Mitra, Sparse channel estimation with zero tap detection. IEEE Trans. Wireless Commun. **6**(5), 1743–1753 (2007)
14. B. Wang, F. Gao, S. Jin, H. Lin, G.Y. Li, Spatial- and frequency wideband effects in millimeter-wave massive MIMO systems. IEEE Trans. Signal Process. **66**(13), 3393–3406 (2018)
15. J. Mo, P. Schniter, R. Heath, Channel estimation in broadband millimeter wave MIMO systems with few-bit ADCs. IEEE Trans. Signal Process. **66**(5), 1141–1154 (2018)
16. A. Alkhateeb, R. Heath, Frequency-selective hybrid precoding for limited feedback millimeter wave systems. IEEE Trans. Commun. **64**(5), 1801–1818 (2016)
17. S. Buzzi, C. Andrea, On clustered statistical MIMO millimeter wave channel simulation (2016, preprint). arXiv:1604.00648
18. S. Gao, X. Cheng, L. Yang, making wideband channel estimation feasible for mmWave massive MIMO: a doubly sparse approach, in *Proc of IEEE International Conference on Communications, Shanghai* (2019)
19. X. Ma, L. Yang, G.B. Giannakis, Optimal training for MIMO frequency-selective fading channels. IEEE Trans. Wireless Commun. **4**(2), 453–466 (2005)
20. X. He, R. Song, W.P. Zhu, Pilot allocation for distributed compressed sensing based sparse channel estimation in MIMO-OFDM systems. IEEE Trans. Veh. Technol. **65**(5), 2990–3004 (2016)
21. Z. Zhang, B.D. Rao, Extension of SBL algorithms for the recovery of block sparse signals with intra-block correlation. IEEE Trans. Signal Process. **61**(8), 2009–2015 (2013)
22. T. Zemen, C.F. Mecklenbrauker, Time-variant channel estimation using discrete prolate spheroidal sequences. IEEE Trans. Signal Process. **53**(9), 3597–3607 (2005)
23. A.B. Awoseyila, C. Kasparis, B.G. Evans, Improved single frequency estimation with wide acquisition range. Electron. Lett. **44**(3), 245–247 (2008)
24. M.A. Davenport, M.B. Wakin, Analysis of orthogonal matching pursuit using the restricted isometry property. IEEE Trans. Inf. Theory **56**(9), 4395–4401 (2010)

Chapter 4
Generic Millimeter-Wave Multi-User Transceiver Design

Abstract This chapter works on crafting a generally enhanced transceiver explicitly for wideband multi-user (wMU) mmWave massive multiple-input multiple-output (mMIMO), aiming to maximize mutual information (MI). The proposed scheme follows the prevalent hybrid block diagonalization (HBD-)based framework and further proves that HBD is optimal in the sense of MI. Specifically, the transceiver design decouples hybrid processing into two-stage analog and digital processing, with which one can derive the MI bounds associated with HBD. After demonstrating the bound tightness, we obtain the excellent HBD transceivers by optimizing the lower bound. The proposed HBD technique does not rely on substantial computational complexity, striking channel sparsity, or high-resolution analog beamformers and can achieve a superb MI performance even with inferior hardware configurations.

Keywords mmWave · Massive multiple-input multiple-output (mMIMO) · Wideband · Multi-user · Hybrid block diagonalization · Mutual information · Transceiver design

4.1 Background

4.1.1 Introduction of Multi-User Massive MIMO

To meet the stringent data requirement in 5G and Beyond, a well-recognized direction is to scale up the existing multi-input multi-output (MIMO) systems from the aspects of operating bandwidth and antenna dimension. With judicious transceiver design, the resultant mmWave massive MIMO (mMIMO) can not only enhance the link quality but also facilitate spatial multiplexing. Owing to these prominent merits, a single mmWave mMIMO base station (BS) is expected to serve multiple user equipments (UEs) simultaneously [1–3]. The resultant wideband multi-user (wMU) mmWave mMIMO systems are expected to pave the way towards massive connection and ultra-fast speed for the next-generation cellular.

X. Cheng et al., *mmWave Massive MIMO Vehicular Communications*,
Wireless Networks, https://doi.org/10.1007/978-3-030-97508-1_4

In general, accurate channel state information (CSI) will be indispensable to mmWave mMIMO transceiver design. Fortunately, a series of proposed channel estimators (e.g., [4, 5]) can help address the CSI issues and let us focus on the specific transceiver design itself. This topic has received increasing attention from both academia and industry in recent years, but the relevant study for wideband multi-user (wMU) scenarios is still far from mature. Compared to the well studied narrowband point-to-point (P2P) counterparts, the mMU case encounters three significant challenges in transceiver design. First, the unique hybrid structure imposes more constraints on the transceiver design [6–8]. Secondly, a shared analog beamformer renders coupled (as opposed to independent) subcarrier processing in orthogonal-frequency-division-multiplexing (OFDM) regime [9]. Finally, the wMU scenarios have to jointly consider the user-specific signal quality variation and the multi-user interference (MUI) [10]. To tackle these challenges, quite a few methods have been proposed in the existing literature. Among these various candidates, a so-termed hybrid block diagonalization (HBD) [11] framework has been commonly adopted for the transceiver design. In brief, HBD will eliminate the MUI at the Tx end with the help of CSI such that detection can be applied individually at the UE end [12].

4.1.2 Design Objectives and Proposed Approach

To gain more insight to HBD, we first survey some of the most representative schemes. The solution proposed in [10] entails analog beam steering and digital MUI cancellation. This one is easy-to-implement so long as a high-resolution analog beamformer is deployed, but its performance is mostly mediocre. Another solution proposed in [13] relies on the so-called equal-gain-transmission (EGT) processing. This technique does not demand a sparse channel, but is a narrowband design and requires the BS to be operated at the multiplexing mode only. Another solution proposed in [14] consists of pre-beam separation and post-subspace projection. This approach works relatively well if the multiplexing gain could perfectly match the rank of the multi-user channel, but its performance is likely compromised once such a condition is violated. In addition, its applicability to wideband channels is also questionable. Although HBD devised in [15] has taken the frequency-selectivity into account, it is essentially a greedy extension of its P2P ancestors, thereby inheriting the original limitations and lacking proper MUI management.

As can be seen, existing techniques fail to offer a performance guarantee for general wMU mmWave mMIMO systems due to their ad-hoc nature. To overcome the present limitations, we seek to design a systematic HBD-based transceiver framework explicitly for wMU mmWave mMIMO. The key steps are as follows:

- By revealing the crucial role of analog processing in HBD-based transceiver design, we transform the coupled hybrid processing into sequential hybrid and digital processing.
- As the angular resolution approaches infinity, we demonstrate the global optimality of HBD in the sense of MI optimization for wMU mmWave mMIMO.
- By optimizing the achievable yet asymptotically optimal MI lower-bound, we end up with the sub-optimal yet high-performance analog processing.
- With the help of the post-digital processing, we accomplish HBD-based transceiver designs for both the multi-aperture structure (MAS) and the multi-beam structure (MBS).

Our judicious design ensures a well-acceptable complexity burden and an excellent performance under various channel environments and hardware configurations.

4.2 System Description and Problem Formulation

This section will showcase the studied system and channel models. Based on the derived input-output relationship, we then formulate the corresponding HBD-based transceiver design problem for wMU mmWave mMIMO.

4.2.1 System and Channel Models

We consider a downlink MU system with its illustrative diagram shown in Fig. 4.1. Specifically, N_t antennas with M_t radio-frequency (RF) chains are employed at the base station (BS). N_r antennas with M_r radio-frequency (RF) chains are employed at each of the K UEs. The BS will transmit N_s data streams to each UE, satisfying

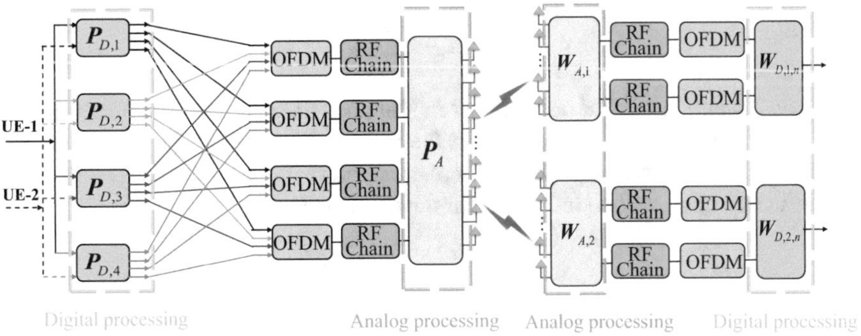

Fig. 4.1 An illustrative wMU mmWave mMIMO system model with $K = 2$, $N = 4$, $M_t = 4$ and $M_r = 2$

$KN_s \leq M_t \leq N_t$ and $N_s \leq M_r \leq N_r$. Without loss of generality, we further assume $M_r = N_s$.

To incorporate the unique characteristics of mmWave channels, a commonly-used geometric wideband channel model is adopted. Let L and D represent the number of dominant paths and delay taps, respectively. Then according to [16], the tap-d channel ($d \in [0, D)$) between the BS and UE-k can be expressed as

$$\mathbf{H}_{k,d} = \sum_{l=1}^{L} \sqrt{\frac{N_t N_r}{L}} \alpha_{l,k} h(dT_s - \tau_{l,k}) \boldsymbol{a}_r(\theta_{l,k}) \boldsymbol{a}_t^*(\phi_{l,k}) \,. \tag{4.1}$$

For path-l, $\alpha_{l,k} \sim \mathcal{CN}(0, 1)$ represents its amplitude; $h(\tau)$ is the pulse-shaping response sampled at τ; $\tau_{l,k}$ is the propagation delay which obeys a uniform distribution over $[0, (N_c - 1)T_s)$; $\theta_{l,k}$ and $\phi_{l,k}$ are the angle of arrival (AoA) and angle of departure (AoD), respectively, both uniformly distributed within $[0, 2\pi)$. With the typical half-wavelength spaced uniform linear array (ULA), the transmit and receive array responses $\boldsymbol{a}_t(\cdot)$ and $\boldsymbol{a}_r(\cdot)$ are respectively given by

$$\boldsymbol{a}_t(\phi) = \frac{1}{\sqrt{N_t}} [1, e^{j\pi \sin\phi}, \cdots, e^{j(N_t - 1)\pi \sin\phi}]^T \,, \tag{4.2a}$$

$$\boldsymbol{a}_r(\theta) = \frac{1}{\sqrt{N_r}} [1, e^{j\pi \sin\theta}, \cdots, e^{j(N_r - 1)\pi \sin\theta}]^T \,. \tag{4.2b}$$

For simplicity, define

$$\boldsymbol{A}_{t,k} = [\boldsymbol{a}_t(\phi_{1,k}), \boldsymbol{a}_t(\phi_{2,k}), \cdots, \boldsymbol{a}_t(\phi_{L,k})] \,, \tag{4.3a}$$

$$\boldsymbol{A}_{r,k} = [\boldsymbol{a}_r(\theta_{1,k}), \boldsymbol{a}_r(\theta_{2,k}), \cdots, \boldsymbol{a}_r(\theta_{L,k})] \,, \tag{4.3b}$$

$$\boldsymbol{\Lambda}_{k,d} = \mathrm{diag}\{\alpha_{l,k} h(dT_s - \tau_{l,k})\}_{l=1}^{L} \,, \tag{4.3c}$$

then $\mathbf{H}_{k,d}$ can be rewritten as

$$\mathbf{H}_{k,d} = \boldsymbol{A}_{r,k} \boldsymbol{\Lambda}_{k,d} \boldsymbol{A}_{t,k}^* \,. \tag{4.4}$$

It is worth mentioning that our channel model does not take Doppler into account. This is because the design focus is on the mutual information. In this sense, the corresponding design is applied to the coherent interval. How to tackle the Doppler effects in detection will be left to the following chapters.

4.2.2 Input-Output Relationship

Let $\boldsymbol{x}_n = \left[\boldsymbol{x}_{1,n}^*, \boldsymbol{x}_{2,n}^*, \cdots, \boldsymbol{x}_{K,n}^*\right]^*$ be the input data on subcarrier-n ($n \in [1, N]$), satisfying $\mathbb{E}\{\boldsymbol{x}_{k,n}\boldsymbol{x}_{k,n}^*\} = \frac{1}{M_r}\boldsymbol{I}_{M_r}$. $\boldsymbol{x}_{k,n}$ will be sent to UE-k ($k \in [1, k]$) and its symbols are selected from a Gaussian constellation. At the BS-end, $\boldsymbol{x}_n$ is first digitally precoded by

$$\boldsymbol{P}_{D,n} = [\boldsymbol{P}_{D,1,n}, \boldsymbol{P}_{D,2,n}, \cdots, \boldsymbol{P}_{D,K,n}] \in \mathcal{C}^{M_t \times KM_r}, \tag{4.5}$$

followed by M_t N-point inverse fast Fourier transform's (IFFT's). Then, the obtained time-domain signal is appended by a length-D ($D \geq N_c - 1$) cyclic-prefix (CP) before being processed by an analog precoder $\boldsymbol{P}_A \in \mathcal{C}^{N_t \times M_t}$. Note that $\boldsymbol{P}_A$ is applied after IFFT in hybrid OFDM systems, thus it remains identical across all subcarriers. This feature drastically differs from conventional fully digital OFDM systems, where the precoders are independent and distinct across subcarriers.

After hybrid precoding, the transmitted signal on subcarrier-n becomes

$$\boldsymbol{s}_n = \boldsymbol{P}_A \boldsymbol{P}_{D,n} \boldsymbol{x}_n . \tag{4.6}$$

Based on the time-domain channel $\mathbf{H}_{k,d}$ in Eq. (4.1), we can obtain its frequency-domain version as

$$\boldsymbol{H}_{k,n} = \sum_{d=0}^{D-1} \mathbf{H}_{k,d} e^{-j\frac{2\pi n}{N}d} . \tag{4.7}$$

As a result, the signal captured by UE-k on subcarrier-n is

$$\boldsymbol{r}_{k,n} = \boldsymbol{H}_{k,n}\boldsymbol{s}_n + \boldsymbol{\eta}_{k,n} , \tag{4.8}$$

with $\boldsymbol{\eta}_{k,n} \sim \mathcal{CN}(\mathbf{0}, \sigma^2 \boldsymbol{I}_{N_r})$ being the Gaussian noise. In this chapter, we define the signal-to-noise ratio (SNR) as $\frac{1}{\sigma^2}$.

After receiving $\boldsymbol{r}_{k,n}$, the analog combiner, denoted as $\boldsymbol{W}_{A,k} \in \mathcal{C}^{N_r \times M_r}$, will be applied. With CP removal and M_r N-point fast Fourier transform's (FFT's), the frequency-domain signal will be digitally combined by $\boldsymbol{W}_{D,k,n} \in \mathcal{C}^{M_r \times M_r}$, resulting in

$$\begin{aligned}\boldsymbol{y}_{k,n} &= \boldsymbol{W}_{D,k,n}^* \boldsymbol{W}_{A,k}^* \boldsymbol{H}_{k,n} \boldsymbol{P}_A \boldsymbol{P}_{D,n} \boldsymbol{x}_n + \boldsymbol{W}_{D,k,n}^* \boldsymbol{W}_{A,k}^* \boldsymbol{\eta}_{k,n} \\ &= \boldsymbol{W}_{D,k,n}^* \widetilde{\boldsymbol{H}}_{k,n} \boldsymbol{P}_{D,n} \boldsymbol{x}_n + \boldsymbol{\xi}_{k,n} ,\end{aligned} \tag{4.9}$$

We term $\widetilde{\boldsymbol{H}}_{k,n} = \boldsymbol{W}_{A,k}^* \boldsymbol{H}_{k,n} \boldsymbol{P}_A$ as the equivalent digital channel (EDC) associated with UE-k on subcarrier-n. By stacking $\{\boldsymbol{y}_{k,n}\}_{n=1}^N$, the system I-O relationship is established as

$$\boldsymbol{y}_n = \boldsymbol{W}^*_{D,n} \underbrace{\begin{bmatrix} \widetilde{\boldsymbol{H}}_{1,n} \\ \widetilde{\boldsymbol{H}}_{2,n} \\ \vdots \\ \widetilde{\boldsymbol{H}}_{K,n} \end{bmatrix}}_{\widetilde{\boldsymbol{H}}_n} \boldsymbol{P}_{D,n}\boldsymbol{x}_n + \boldsymbol{\xi}_n \ , \tag{4.10}$$

with $\boldsymbol{W}_{D,n} = \text{diag}\{\boldsymbol{W}_{D,k,n}\}_{k=1}^K$, $\boldsymbol{\xi}_n$ being the concatenated noise vector, and $\widetilde{\boldsymbol{H}}_n$ being the multi-user EDC on subcarrier-n. It is worth mentioning that $\boldsymbol{W}_{D,n}$ and $\boldsymbol{P}_{D,n}$ in Eq. (4.10) is associated with the digital part, while $\widetilde{\boldsymbol{H}}_n$ associated with analog part.

4.2.3 Problem Formulation

Our aim is to develop a systematic HBD scheme for wMU systems. More specifically, with known $\mathbf{H}_{k,d}$, determine the digital part $\boldsymbol{P}_{D,n}$ and $\boldsymbol{W}_{D,k,n}$, as well as the analog part $\boldsymbol{P}_A$ and $\boldsymbol{W}_{A,k}$, to maximize the system MI while satisfying the following prerequisite for HBD:

$$\widetilde{\boldsymbol{H}}_n \boldsymbol{P}_D = \text{diag}\left\{\widetilde{\boldsymbol{H}}_n[(k-1)M_r + 1 : kM_r, :]\boldsymbol{P}_{D,k,n}\right\}_{k=1}^K \ . \tag{4.11}$$

Once each UE enjoys an MUI-free EDC, the achievable MI for UE-k on subcarrier-n is computed as

$$I_{k,n} = \det\left(\boldsymbol{I}_{M_r} + \frac{1}{M_r\sigma^2}\left(\boldsymbol{W}^*_{D,k,n}\boldsymbol{W}^*_{A,k}\boldsymbol{W}_{A,k}\boldsymbol{W}_{D,k,n}\right)^{-1} \times \boldsymbol{H}_{e,k,n}\boldsymbol{H}^*_{e,k,n}\right) , \tag{4.12}$$

with $\boldsymbol{H}_{e,k,n} = \boldsymbol{W}^*_{D,k,n}\boldsymbol{W}^*_{A,k}\boldsymbol{H}_{k,n}\boldsymbol{P}_A\boldsymbol{P}_{D,k,n}$. As a result, the HBD-based transceiver design problem for wMU mmWave mMIMO can be formulated as:

P1. Problem Statement 1 [*Generic HBD-wMU transceiver design*]

$$\underset{\boldsymbol{W}_{D,k,n},\boldsymbol{W}_{A,k},\boldsymbol{P}_A,\boldsymbol{P}_{D,n}}{\text{arg max}} \quad I = \sum_{k=1}^{K}\sum_{n=1}^{N} \log_2 I_{k,n} \ , \tag{4.13a}$$

$$\text{subject to} \quad Eq. (4.11) \ , \tag{4.13b}$$

$$\boldsymbol{P}_A \in \mathcal{F} \ , \tag{4.13c}$$

$$\forall k, \quad \boldsymbol{W}_{A,k} \in \mathcal{W} \ , \tag{4.13d}$$

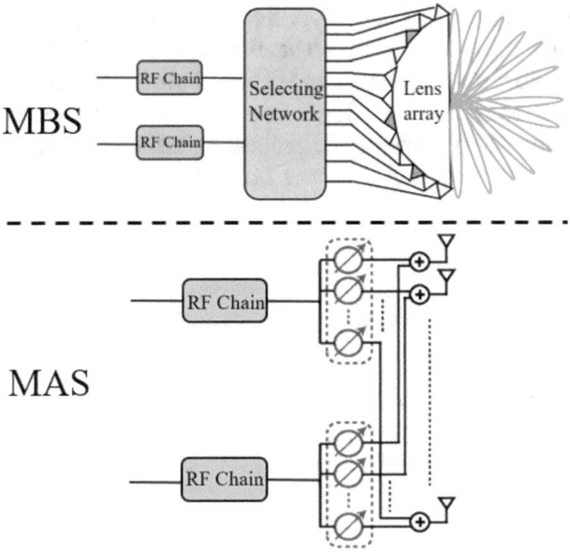

Fig. 4.2 The diagram of MBS and MAS

$$\forall k, n, \parallel \boldsymbol{P}_A \boldsymbol{P}_{D,k,n} \parallel \leq M_r \ . \tag{4.13e}$$

In **P1**, $\mathcal{F}$ and $\mathcal{W}$ stand for the feasible sets of analog precoders,[1] which exhibit different constant-modulus forms in two mainstream hybrid mMIMO structures, namely MBS (shown in the upper part of Fig. 4.2) and MAS (shown in the lower part of Fig. 4.2). The former corresponds to the so-called lens-array antennas [17], while the latter fits to the fully-connected analog phase shifter (APS) network [18].

4.2.4 *Design Strategy*

A common practice for obtaining the optimal precoders is by directly solving the formulated optimization problem **P1**. However, the straightforward manner turns out to be intractable because of no explicit objective function of **P1**. Nevertheless, by observing Eq. (4.11), we find that the BS-end digital precoder $\boldsymbol{P}_{D,k,n}$ resides in the null-space of $\widetilde{\boldsymbol{H}}_n$, indicating that the former is determined by the latter. If Eq. (4.11) holds, UE-k will enjoy an MUI-free EDC $\boldsymbol{W}^*_{A,k}\boldsymbol{H}_{k,n}\boldsymbol{P}_A\boldsymbol{P}_{D,n}$, which is a function of $\widetilde{\boldsymbol{H}}_n$. A similar conclusion can be drawn for $\boldsymbol{W}_{D,k,n}$ as well. In consequence, the digital part (including $\boldsymbol{P}_{D,n}$ and $\boldsymbol{W}_{D,k,n}$) is essentially a function of the analog part

[1] We do not deliberately distinguish between the precoder and combiner, so both terms will be named precoder without ambiguity.

(including $\boldsymbol{W}_{A,k}$ and $\boldsymbol{P}_A$). Such a property makes it valid to transform the joint processing design into sequential (analog and digital) processing.

To be more specific, an MI bound can be derived in replacement of the inexplicit I once giving the analog part, so the analog precoders can be determined via bound optimization. The digital processing can then be complemented to meet the HBD condition whist maximizing MI. We then start to detail the HBD-based transceiver design.

4.3 Mutual Information (MI) Bounds

In this section, we first derive the MI upper- and lower-bounds via HBD. Then, we will investigate the derived bounds and reveal the global optimality of HBD in the sense of MI optimization in wMU mmWave mMIMO transceiver design.

4.3.1 MI Upper-Bound

For simplicity, we express the system I-O relationship in a concise form as

$$\boldsymbol{y} = \underbrace{\mathrm{diag}\{\boldsymbol{W}_{D,n}^{*}\}_{n=1}^{N}}_{\boldsymbol{W}_D^*} \underbrace{\mathrm{diag}\{\widetilde{\boldsymbol{H}}_n\}_{n=1}^{N}}_{\widetilde{\boldsymbol{H}}} \underbrace{\mathrm{diag}\{\boldsymbol{P}_{D,n}\}_{n=1}^{N}}_{\boldsymbol{P}_D} \boldsymbol{x} + \boldsymbol{\xi} \,. \tag{4.14}$$

$\boldsymbol{W}_{D,n}$ bears a block-diagonal form in a multi-user setup, but this restriction could be relaxed if cooperation is allowed among UEs. If this is the case, let us perform singular value decomposition (SVD) for $\widetilde{\boldsymbol{H}}_n$, i.e., $svd(\widetilde{\boldsymbol{H}}_n) = \widetilde{\boldsymbol{U}}_n \widetilde{\boldsymbol{\Sigma}}_n \widetilde{\boldsymbol{V}}_n^*$. Then setting $\boldsymbol{P}_{D,n} = \widetilde{\boldsymbol{V}}_n$ and $\boldsymbol{W}_{D,n} = \widetilde{\boldsymbol{U}}_n$ leads $\boldsymbol{P}_D$ and $\boldsymbol{W}_D$ to the right and left singular matrices of $\widetilde{\boldsymbol{H}}$. The corresponding MI can therefore be computed as

$$\begin{aligned} I_U &= \sum_{n=1}^{N} \sum_{i=1}^{KM_r} \log_2 \left(1 + \frac{\widetilde{\Sigma}_n^2[i,i]}{M_r \sigma^2}\right) \\ &= \sum_{i=1}^{NKM_r} \log_2 \left(1 + \frac{\widetilde{\Sigma}^2[i,i]}{M_r \sigma^2}\right) , \end{aligned} \tag{4.15}$$

with $\widetilde{\Sigma}[i,i]$ representing the i-th eigenvalue of $\widetilde{\boldsymbol{H}}$. Since Eq. (4.15) is obtained through the SVD of $\widetilde{\boldsymbol{H}}$ under user cooperation, it is clear that I_U is an MI upper-bound for HBD-based wMU transceivers.

4.3.2 MI Lower-Bound

If without cooperation among UEs, $\boldsymbol{W}_{D,n} = \boldsymbol{I}$ and $\boldsymbol{P}_{D,n} = \widetilde{\boldsymbol{H}}_n^{\dagger}\text{diag}\{\|\widetilde{\boldsymbol{H}}_n^{\dagger}[:,i]\|_F^{-1}\}_{i=1}^{NM_r}$. It can be verified that $\boldsymbol{P}_D$ acts as the zero-forcing (ZF) pre-equalizer, with which the system MI is

$$\begin{aligned} I_L &= \sum_{n=1}^{N}\sum_{i=1}^{KM_r}\log_2\left(1+\frac{\|\widetilde{\boldsymbol{H}}_n^{\dagger}[:,i]\|_F^{-2}}{M_r\sigma^2}\right) \\ &= \sum_{i=1}^{NKM_r}\log_2\left(1+\frac{\|\widetilde{\boldsymbol{H}}^{\dagger}[:,i]\|_F^{-2}}{M_r\sigma^2}\right). \end{aligned} \tag{4.16}$$

In the above, the BS-end processing not only cancels out the MUI but also the inter-stream interference for an individual UE. Hence this form is essentially a special type of HBD. Besides, achieving I_L does not involve any UE-end digital processing, so I_L behaves as the MI lower-bound.

4.3.3 MI Relationship

Typically, I_L will be strictly smaller than I_U. While as the number of antennas (N_t, N_r) approaches infinity, these two bounds will merge with probability 1. To verify this point, first rewrite Eq. (4.7) as

$$\boldsymbol{H}_{k,n} = \boldsymbol{A}_{r,k}\left(\sum_{d=0}^{D-1}\boldsymbol{\Lambda}_{k,d}e^{-j\frac{2\pi n}{N}d}\right)\boldsymbol{A}_{t,k}^* . \tag{4.17}$$

In the case of infinite N_t and N_r, we have

$$\boldsymbol{a}_{t(r)}^*(\theta_1)\boldsymbol{a}_{t(r)}(\theta_2) = \begin{cases} 0 & \theta_1 \neq \theta_2 \\ 1 & \theta_1 = \theta_2 \end{cases}. \tag{4.18}$$

Both $\boldsymbol{A}_{t,k}$ and $\boldsymbol{A}_{r,k}$ are semi-unitary matrices. Meanwhile, the term inside the bracket of Eq. (4.17) is a diagonal matrix, so $\forall n$, $\{\boldsymbol{H}_{k,n}\}_{n=1}^{N}$ share common singular vectors. Utilizing this property, the HBD-based transceivers as N_t and N_r go to infinity can be designed as

$$\boldsymbol{P}_A = \left[\overline{\boldsymbol{A}}_{t,1}, \overline{\boldsymbol{A}}_{t,2}, \cdots, \overline{\boldsymbol{A}}_{t,K}, \overline{\boldsymbol{A}}\right] , \tag{4.19a}$$

$$\boldsymbol{W}_{A,k} = \overline{\boldsymbol{A}}_{r,k} , \tag{4.19b}$$

$$\boldsymbol{P}_{D,n} = \left[\boldsymbol{I}_{KM_r}^*, \boldsymbol{0}_{M_t-KM_r}^*\right]^* , \tag{4.19c}$$

$$\boldsymbol{W}_{D,k,n} = \boldsymbol{I}_{M_r} . \tag{4.19d}$$

Specifically, $\overline{\boldsymbol{A}}_{t,k} \in \mathcal{C}^{N_t \times M_r}$ is an arbitrary right singular sub-matrix extracted from $\boldsymbol{A}_{t,k}$, $\overline{\boldsymbol{A}}_{r,k}$ is the corresponding left singular sub-matrix extracted from $\boldsymbol{A}_{r,k}$, and $\overline{\boldsymbol{A}}$ is an arbitrary matrix bearing constant-modulus form. This is a valid design because all the analog precoders are constant-modulus, and $\| \boldsymbol{P}_A \boldsymbol{P}_{D,n} \|_F = M_r$ satisfies Eq. (4.13e), i.e., the transmit power constraint. The above setting leads $\widetilde{\boldsymbol{H}}$ defined in Eq. (4.14) to be a diagonal matrix, hence I_U achieved via SVD equals to I_L achieved via ZF pre-equalization.

4.3.4 HBD Optimality

Although the proposed HBD solution reveals that $I_L = I_U$ as $N_t, N_r \to \infty$, a general optimality cannot be guaranteed because $\overline{\boldsymbol{A}}_{r,k}$ (or $\overline{\boldsymbol{A}}_{t,k}$) in Eq. (4.19) has multiple choices, thereby awaiting further determination.

Note that, regardless of $\overline{\boldsymbol{A}}_{r,k}$, the above design always assures that $\boldsymbol{H}_{k,n}$ (see Eq. (4.17)) can be decomposed into M_r parallel single-input single-output (SISO) channels, each of which stands for a singular vector pair (SVP). Suppose that SVP-i_k ($i_k \leq L$) is selected for UE-k, then the MI contributed by this one can be individually computed as

$$I_{k,i_k} = \sum_{n=1}^{N} \log_2 \left(1 + \frac{|\sum_{d=0}^{D-1} \Lambda_{k,d}[i_k, i_k] e^{-j\frac{2\pi n}{N}d}|^2}{M_r \sigma^2}\right). \tag{4.20}$$

To maximize I_U or I_L, choosing M_r out of L SVPs with the most significant MI contribution would suffice. $\overline{\boldsymbol{A}}_{t,k}$ and $\overline{\boldsymbol{A}}_{r,k}$ are automatically determined. Interestingly, with this method, the ultimately achieved MI not only reaches the optimum among HBD-based wMU transceivers, but also among *any* wMU transceivers. For clarity, a proposition is expressly provided as follows.

Proposition 1 (Optimality of HBD) *With an infinite number of antennas employed at the transceivers, HBD is the optimal precoding technique in the sense of MI optimization for wMU mmWave mMIMO transceiver design.*

Proof In P2P scenarios, it has been proven that the near-optimal analog precoder is the Karcher mean of the optimal unconstrained precoders on individual subcarrier (see Eqs. (26) and (27) in [9], where "≈" is exactly "=" when $N_t, N_r \to \infty$, thus the near optimality becomes exact optimality). For UE-k, the optimal unconstrained precoder on each subcarrier is a sub-matrix of $\boldsymbol{A}_{t,k}$, hence the resultant Karcher mean is still a sub-matrix of $\boldsymbol{A}_{t,k}$. A similar conclusion holds for $\boldsymbol{A}_{r,k}$ as well. Therefore, choosing the best M_r out of L SVPs to construct the analog precoder for UE-k clearly leads to optimality. Now that all UEs enjoy a MUI-free EDC, individual optimality automatically generates overall optimality.

4.4 Transceiver Design

In this section, we will develop HBD-based transceivers for practical wMU mmWave mMIMO systems. The first step is to determine the analog processing by optimizing the MI bound. The second step is to complete the digital processing to reach MUI-free target and improve individual MI.

4.4.1 Analog-Domain Processing

Since the MI lower-bound I_L is always achievable and asymptotically optimal in the limit case, it is reasonable to use it as a replacement of the inexplicit I. For simplicity, we make a slight modification to I_L by omitting the "1" term in Eq. (4.16), giving rise to

$$\begin{aligned} I_L &= \sum_{i=1}^{NKM_r} \log_2\left(1+\frac{\|\widetilde{\boldsymbol{H}}^{\dagger}[:,i]\|_F^{-2}}{M_r\sigma^2}\right) \\ &\approx -\log_2\left\{\prod_{i=1}^{NKM_r}(\widetilde{\boldsymbol{H}}\widetilde{\boldsymbol{H}}^*)^{-1}[i,i]\right\} - NKM_r\log_2 M_r\sigma^2 . \end{aligned} \tag{4.21}$$

Equation (4.21) monotonically decreases with the term inside the bracket, regardless of SNR. This makes analog processing a following optimization problem:

P2. Problem Statement 2 [*Analog processing via optimizing* I_L]

$$\underset{\boldsymbol{W}_{A,k},\boldsymbol{P}_A}{\arg\min} \quad \prod_{i=1}^{NKM_r}(\widetilde{\boldsymbol{H}}\widetilde{\boldsymbol{H}}^*)^{-1}[i,i] , \tag{4.22a}$$

$$\text{subject to} \quad \boldsymbol{P}_A \in \mathcal{F} , \quad \forall k, \boldsymbol{W}_{A,k} \in \mathcal{W} . \tag{4.22b}$$

As has been mentioned before, there are two mainstream hybrid mmWave mMIMO structures, namely MBS and MAS. Since **P2** does not use any specialties in terms of the analog structures, a similar design paradigm can be shared by both after taking their individual constraints into account.

4.4.1.1 MBS

Denote $\mathcal{F}_M$ as the set containing all M-dimensional DFT base. The analog constraints for MBS can be expressed as

$$\forall n, k, \boldsymbol{P}_A[:, n] \in \mathcal{F}_{N_t}, \ \boldsymbol{W}_{A,k}[:, n] \in \mathcal{F}_{N_r}. \tag{4.23}$$

The minimizer of **P2** can be acquired via $\mathbb{C}_{N_t}^{M_t}\left(\mathbb{C}_{N_r}^{M_r}\right)^K$ trials, but the complexity is obviously prohibitive under the wMU mMIMO setup (>10^{14} trials for $N_r = 16$, $N_t = 64$, $M_r = 2$, $M_t = 6$, $K = 3$). A natural means to reduce complexity is by shrinking the search space as in [8], whereas two drawbacks will emerge here. First, the effectiveness is highly sensitive to the sparsity level of mmWave channels. Secondly, even if an extreme sparsity holds, the complexity is still growing *exponentially* with the number of UEs. In light of these deficiencies, we judiciously devise a double-sequential-search (DSS) algorithm for MBS. The pseudo-code is provided in Algorithm 2.

DSS algorithm entails greedy codeword selection for constructing the UE-end analog precoders and greedy codeword exclusion for constructing the BS-end analog precoder. We start from the UE-end by setting $\boldsymbol{P}_A = \boldsymbol{F}_{N_t}$, which can remove the BS-end influence. Utilizing the notation defined in **Step.1**, codeword selection at the inner-loop (m, k) (**Step.2–Step.7**) can be implemented as

$$m^* = \arg\min_{m} \prod_{i=1}^{NKM_r} \left(\widetilde{\boldsymbol{H}}(m,k)\widetilde{\boldsymbol{H}}^*(m,k)\right)^{-1}[i,i], \tag{4.24}$$

where $\widetilde{\boldsymbol{H}}(m,k) = \text{diag}\left\{\widetilde{\boldsymbol{H}}_n(m,k)\right\}_{n=1}^{N}$ satisfying

Algorithm 2 DSS analog processing algorithm for the MBS

Require: $\forall k, n, \boldsymbol{H}_{k,n}, \boldsymbol{F}_{N_t}, \boldsymbol{F}_{N_r}, K, M_r$ and M_t;
Ensure: $\forall k, \boldsymbol{W}_{A,k}, \boldsymbol{P}_A$
1: **Initialization** $\forall k, \widetilde{\boldsymbol{W}}_{R,k} = \varnothing, \widetilde{\boldsymbol{P}}_A = \boldsymbol{F}_{N_t}, count = N_t, \mathcal{I} = \{1, 2, \cdots, N_t\}$;
2: **for** $m \le M_r$ **do**
3:     **for** $k \le K$ **do**
4:         *Compute Eq. (4.24) to get* m^* ;
5:         $\widetilde{\boldsymbol{W}}_{A,k} = \left[\widetilde{\boldsymbol{W}}_{A,k}, \boldsymbol{F}_{N_r}[:, m^*]\right]$;
6:     **end for**
7: **end for**
8: **while** $count > M_t$ **do**
9:     *Compute Eq. (4.27) to get* j^* ;
10:     $\mathcal{I} = \mathcal{I} \backslash \mathcal{I}(j^*)$;
11:     $\widetilde{\boldsymbol{P}}_A = \boldsymbol{F}_{N_t}[:, \mathcal{I}]$;
12:     $count = count - 1$;
13: **end while**
14: $\forall k, \boldsymbol{W}_{A,k} = \widetilde{\boldsymbol{W}}_{R,k}, \boldsymbol{P}_A = \widetilde{\boldsymbol{P}}_A$;

$$\widetilde{\boldsymbol{H}}_n(m,k) = \begin{bmatrix} \widetilde{\boldsymbol{W}}^*_{R,1}\boldsymbol{H}_{1,n} \\ \vdots \\ \widetilde{\boldsymbol{W}}^*_{R,k-1}\boldsymbol{H}_{k-1,n} \\ [\widetilde{\boldsymbol{W}}_{R,k}, \boldsymbol{F}_{N_r}[:,m]]^*\boldsymbol{H}_{k,n} \\ \widetilde{\boldsymbol{W}}^*_{R,k+1}\boldsymbol{H}_{k+1,n} \\ \vdots \\ \widetilde{\boldsymbol{W}}^*_{R,K}\boldsymbol{H}_{K,n} \end{bmatrix} \boldsymbol{F}_{N_t} . \tag{4.25}$$

This criterion ensures a linear complexity with the UE number and maintain the UEs' fairness. After getting $\boldsymbol{W}_{A,k}$, the hybrid structure forces $\boldsymbol{P}_A$ to delete redundant codewords so that only M_t codewords retain. To this end, codeword exclusion at the loop-j (**Step**.8–**Step**.13) can be implemented as

$$j^* = \arg\min_j \prod_{i=1}^{NKM_r} \left(\widetilde{\boldsymbol{H}}(j)\widetilde{\boldsymbol{H}}^*(j)\right)^{-1}[i,i] , \tag{4.26}$$

where $\widetilde{\boldsymbol{H}}(j) = \text{diag}\{\widetilde{\boldsymbol{H}}_n(j)\}_{n=1}^N$ with

$$\widetilde{\boldsymbol{H}}_n(j) = \begin{bmatrix} \widetilde{\boldsymbol{W}}^*_{R,1}\boldsymbol{H}_{1,n} \\ \widetilde{\boldsymbol{W}}^*_{R,2}\boldsymbol{H}_{2,n} \\ \vdots \\ \widetilde{\boldsymbol{W}}^*_{R,K}\boldsymbol{H}_{K,n} \end{bmatrix} \boldsymbol{F}_{N_t}[:, \mathcal{I}\backslash\mathcal{I}(i)] . \tag{4.27}$$

In each iteration, one codeword making the least contribution to system MI will be excluded until M_t codewords are left.

4.4.1.2 MAS

Assume that the MAS consists of b-bit APS, whose adjustable angles are given by

$$\mathcal{B} = \left\{0, 2\pi/2^b, \cdots, 2\pi \times (2^b-1)/2^b\right\} . \tag{4.28}$$

The analog constraints in the MAS are imposed on each entry, giving rise to

$$\forall m,n,k,\ \boldsymbol{P}_A[m,n] \in \frac{e^{j\mathcal{B}}}{\sqrt{N_t}},\ \boldsymbol{W}_{A,k}[m,n] \in \frac{e^{j\mathcal{B}}}{\sqrt{N_r}} . \tag{4.29}$$

P2 is a massive NP-hard problem, whose minimizer is difficult to be obtained in a brute force manner. Furthermore, DSS algorithm developed for MBS is also inapplicable here because MAS does not obey a column-wise constraint. To secure a local minimizer within an acceptable complexity, we resort to the entry-wise update.

We first apply DSS algorithm and quantize its output, i.e., $\boldsymbol{P}_A^{(MBS)}$ and $\boldsymbol{W}_{A,k}^{(MBS)}$, as

$$\boldsymbol{P}_R = \frac{1}{\sqrt{N_t}} e^{jQ(\angle \boldsymbol{P}_R^{(MBS)})}, \tag{4.30a}$$

$$\boldsymbol{W}_{A,k} = \frac{1}{\sqrt{N_r}} e^{jQ(\angle \boldsymbol{W}_{A,k}^{(MBS)})}, \tag{4.30b}$$

with $Q(x) = \mathcal{B}\big(\arg\min_i \bmod (x - \mathcal{B}(i), 2\pi)\big)$. $\boldsymbol{P}_R$ and $\boldsymbol{W}_{A,k}$ are expected to be a decent initializer. We then take $\boldsymbol{P}_A[a, b]$ to exemplify the entry-wise update by fixing the rest elements of $\boldsymbol{P}_A$ and $\{\boldsymbol{W}_{A,k}\}_{k=1}^K$. By testing all candidates within $\frac{1}{\sqrt{N_t}} e^{j\mathcal{B}}$, the one minimizing Eq. (4.22a) will be used for replacing the current $\boldsymbol{P}_A[a, b]$. If the entire $\boldsymbol{P}_A$ is refreshed, a similar operation is applied to $\{\boldsymbol{W}_{A,k}\}_{k=1}^K$. The update keeps running until triggering a specific terminating condition.

4.4.1.3 Subcarrier Down-Sampling

For either MAS or MBS, analog processing complexity grows linearly with the subcarrier number, which is generally huge in mmWave systems. For this reason, we will activate part of the subcarriers to reduce complexity burden in analog processing. Specifically, we first r-time down-sampling to the subcarriers, thus the original multi-user EDC $\mathrm{diag}\{\widetilde{\boldsymbol{H}}\}_{n=0}^N$ can be replaced by

$$\widetilde{\boldsymbol{H}}_{DS,r} = \mathrm{diag}\{\widetilde{\boldsymbol{H}}_{1+nr}\}_{n=0}^{N/r-1}. \tag{4.31}$$

Consequently, the objective function in **P2** will be adapted to

$$\prod_{i=1}^{NKM_r/r} (\widetilde{\boldsymbol{H}}_{DS,r} \widetilde{\boldsymbol{H}}_{DS,r}^*)^{-1}[i, i]. \tag{4.32}$$

The computational complexity will be reduced by r times. Later on, the simulations will show that even a relatively large r causes a negligible MI loss.

4.4.2 Digital-Domain Processing

After determining the analog part, we then complete digital processing targeting at removing MUI and maximizing MI. Without loss of generality, we take UE-k on subcarrier-n as an example. To highlight the twofold functionality of digital processing, we specially decompose the associated digital precoder into

$$\boldsymbol{P}_{D,k,n} = \boldsymbol{P}_{D,k,1,n}\boldsymbol{P}_{D,k,2,n} \ . \tag{4.33}$$

$\boldsymbol{P}_{D,k,1,n} \in C^{M_t \times M_r}$ is to remove MUI, and $\boldsymbol{P}_{D,k,2,n} \in C^{M_r \times M_r}$ is to maximize the individual MI.

4.4.2.1 First-Step Digital-processing

In order to get an MUI-free EDC, Eq. (4.11) points out that $\boldsymbol{P}_{D,k,1,n}$ must lie in the null-space (here referring to column space) of

$$\overline{\boldsymbol{H}}_{k,n} = \left[\widetilde{\boldsymbol{H}}^*_{1,n}, \cdots, \widetilde{\boldsymbol{H}}^*_{k-1,n}, \widetilde{\boldsymbol{H}}^*_{k+1,n}, \cdots, \widetilde{\boldsymbol{H}}^*_{K,n}\right]^* . \tag{4.34}$$

Define $svd(\overline{\boldsymbol{H}}_{k,n}) = \overline{\boldsymbol{U}}_{k,n}\overline{\boldsymbol{\Sigma}}_{k,n}\overline{\boldsymbol{V}}^*_{k,n}$. The null-space of $\overline{\boldsymbol{H}}_{k,n}$ can be extracted as

$$\widehat{\boldsymbol{V}}_{k,n} = \boldsymbol{V}_{k,n}[:, (K-1)M_r + 1 : M_t]. \tag{4.35}$$

It is worth mentioning that $rank(\widehat{\boldsymbol{V}}_{k,n}) \geq M_r$ must hold, otherwise the HBD constraint cannot be satisfied anyhow.

If $M_t = KM_r$, $\boldsymbol{P}_{D,k,1,n} = \widehat{\boldsymbol{V}}_{k,n}$ is clearly the only viable option [19]. Whereas $M_t > KM_r$, $\boldsymbol{P}_{D,k,1,n}$ is not unique because any M_r-dimensional subspace of $\widehat{\boldsymbol{V}}_{k,n}$ would suffice. Nevertheless, randomly selecting a subspace may compromise EDC, especially when M_t and KM_r are considerably different. To harness power and eliminate MUI, we adopt the subspace projection method similar to [20]. Specifically, we denote the MUI-free space and signal space of $\boldsymbol{H}_{k,n}$ as $\boldsymbol{P}^n_{k,n}$ and $\boldsymbol{P}^s_{k,n}$, represented as:

$$\boldsymbol{P}^n_{k,n} = \boldsymbol{I} - \overline{\boldsymbol{H}}^*_{k,n}(\overline{\boldsymbol{H}}_{k,n}\overline{\boldsymbol{H}}^*_{k,n})^{-1}\overline{\boldsymbol{H}}_{k,n} \ , \tag{4.36a}$$

$$\boldsymbol{P}^s_{k,n} = \widetilde{\boldsymbol{H}}^*_{k,n}(\widetilde{\boldsymbol{H}}_{k,n}\widetilde{\boldsymbol{H}}^*_{k,n})^{-1}\widetilde{\boldsymbol{H}}_{k,n} \ . \tag{4.36b}$$

Define $svd(\boldsymbol{P}^s_{k,n}\boldsymbol{P}^n_{k,n}) = \overline{\boldsymbol{U}}_{k,n}\overline{\boldsymbol{\Sigma}}_{k,n}\overline{\boldsymbol{V}}^*_{k,n}$, then $\boldsymbol{P}_{D,k,1,n}$ can be set as $\overline{\boldsymbol{V}}_{k,n}[:, 1 : M_r]$.

4.4.2.2 Second-Step Digital-processing

With $\boldsymbol{P}_{D,k,1,n}$ and the analog precoders, UE-k ends up with an MUI-free EDC on subcarrier-n as

$$\widetilde{\boldsymbol{H}}_{eff,k,n} = \widetilde{\boldsymbol{H}}_{k,n}\boldsymbol{P}_{D,k,1,n} \ , \tag{4.37}$$

The received signal can, therefore, be expressed as

$$y_{k,n} = \widetilde{H}_{eff,k,n} x_{k,n} + \xi_{k,n} \,. \tag{4.38}$$

4.5 Simulations

This section showcases simulations to compare the proposed HBD scheme over other HBD alternatives. Those parameters relating to the mmWave transceiver are set as: $N_t = 32$, $N_t = 16$, $M_t = 8$, $M_t = 2$, and $K = 4$; whose the parameters related to the channel are set as: $D = 8$ and $N = 32$; $h(\cdot)$ is set as the raised-cosine filter with a roll-off factor $\beta = 0.2$. For MAS, it will adopt 3-bit APS, and its entry-wise update for analog processing will terminate either the complete iteration reaches to 5 or the relative changing ratio is smaller than 0.01. The ideal MI upper-bound is achieved in fully-digital structures without considering MUI. All the curves are the average of 500 independent channel realizations.

4.5.1 *MI in Frequency-Selective Channels*

We first compare MI performances in wideband mmWave channels. For BS-HBD [10], the beam codewords are selected from the DFT matrix. For EGT-HBD [13], 5-bit APS is used at the BS and 4-bit APS is used at the UE. For IMD-HBD proposed in [14], continuous APS is used at both ends. For AM-HBD [15], the analog configuration is identical to that of EGT-HBD. Figure 4.3 shows that if the wideband mmWave channels exhibit strong sparsity ($L = 6$ and $L = 12$), the proposed HBD is remarkably superior to others with a <3 dB performance gap to the ideal upper bound. If $L = 100$, all schemes suffer from big performance degradation. This is because the limitations of hybrid structures restricts the power concentration in such a rich scattering environment. Even so, the proposed HBD still outperforms other HBD candidates.

With 4-times down-sampling applied, the resultant MI loss is very minimal for $L = 6$ and 12. Only until $L = 100$ will a small degradation (<0.8 dB) occur because the correlation between two distant subcarriers diminishes when the channel tends to be Rayleigh-fading. These results reveal that subcarrier down-sampling is effective to reduce implementing complexity.

4.5.2 *MI Versus APS Resolution*

To verify the effectiveness of entry-wise update for MAS, we fix $L = 6$ and plot the relative changing ratio versus iteration times in Fig. 4.4. The result shows three

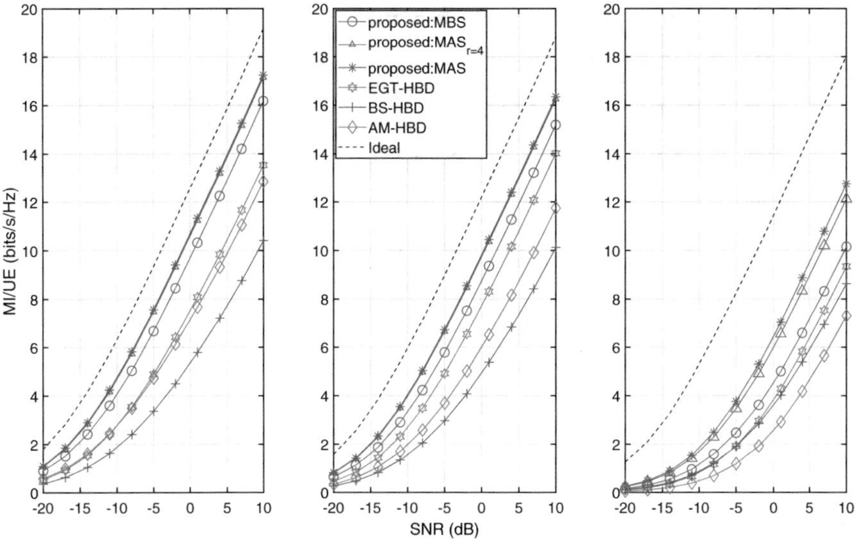

Fig. 4.3 Averaged MI/UE for different HBD schemes in wideband channels with $L = 6$, 12 and 100

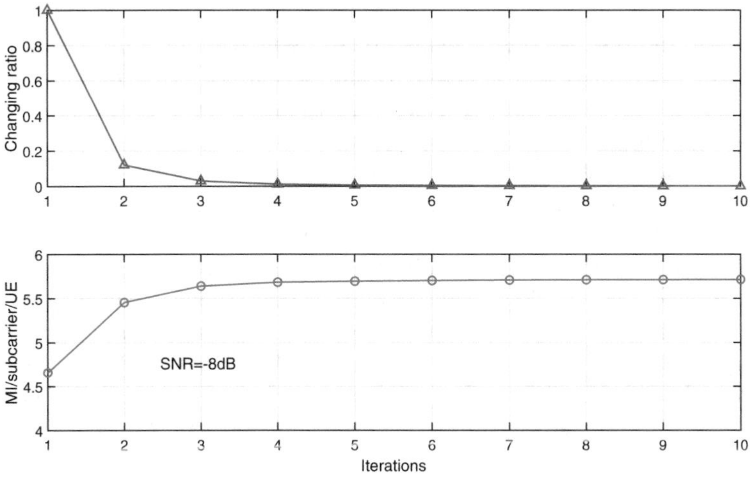

Fig. 4.4 Averaged MI in MAS versus iterations when SNR=-8 dB

iterations suffice to reach a local optimum, implying that the computational burden will not be heavy. By fixing SNR = -8 dB, we see that three iterations harvest an over 20% MI improvement, and 3-bit APS is sufficient for practical use as indicated in Fig. 4.5.

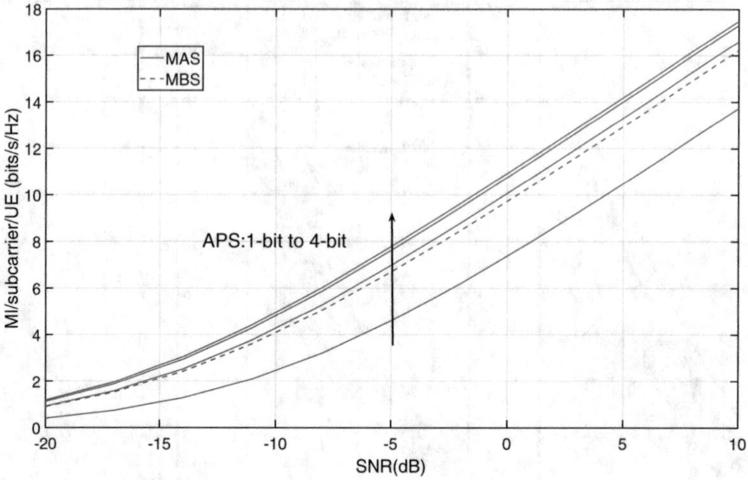

Fig. 4.5 Averaged MI in MAS versus the APS resolution

4.5.3 MI Versus RF Chains

In practice, the BS may utilize more RF chains to strengthen the signal quality (diversity gain). Since this operation mode can be supported by the proposed HBD and AM-HBD, we then take the MBS as an example to compare their MI performance versus the number of RF chains at the BS in Fig. 4.6.

As expected, a dramatic MI improvement is gained from $M_t = 8$ to $M_t = 20$. The principal reason is the coarse beam resolution is partly redeemed by the affluence of RF chains. However, such a positive benefit will saturate rapidly in a sparse channel environment, so a minimal MI improvement can be observed from $M_t = 20$ to $M_t = 32$. The channel power is more dispersively distributed when $L = 100$, thus more RF chains will concentrate the energy and result in a higher MI. Under the same M_t, MI achieved via the proposed HBD is dramatically higher than that of AM-HBD. This is of no surprise because the latter heavily relies on the strong correlation among subcarriers, but the correlation is very weak in large L.

4.5.4 MI Versus UEs and Antennas

By fixing 32 RF chains at the BS, we then simulate the system MI under different UE numbers. Figure 4.7 shows that system MI increases with K in a roughly linear fashion at first. As K becomes larger, the adverse effect arising from the MUI gradually emerges, as reflected by a flatter slope. As K further increases, an MI decline takes place because too many UEs lower the freedom in constructing the MUI-free space, thus compromising EDC's quality.

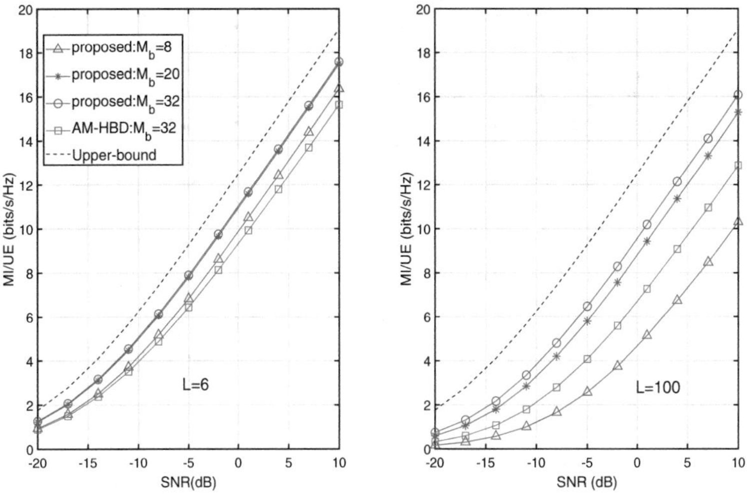

Fig. 4.6 Averaged MI versus the number of BS-end RF chains. AM-HBD is with 5-bit APS at the BS and 4-bit APS at the UE

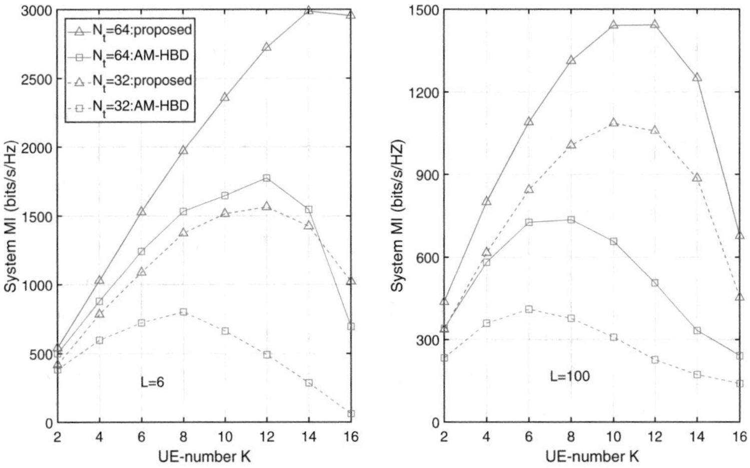

Fig. 4.7 System MI versus the number of UE with $M_t = 32$ and SNR=-8dB. AM-HBD is with 5-bit APS at the BS and 4-bit APS at the UE

By increasing N_t from 32 to 64, we see a delayed appearance of the inflection point where MI starts to decline. The reason is that more antennas augment the beam resolution, thereby easing UEs' separation. With infinite antennas deployed, the beam resolution becomes infinite, so we can expect that the MI curve will monotonically increase with K.

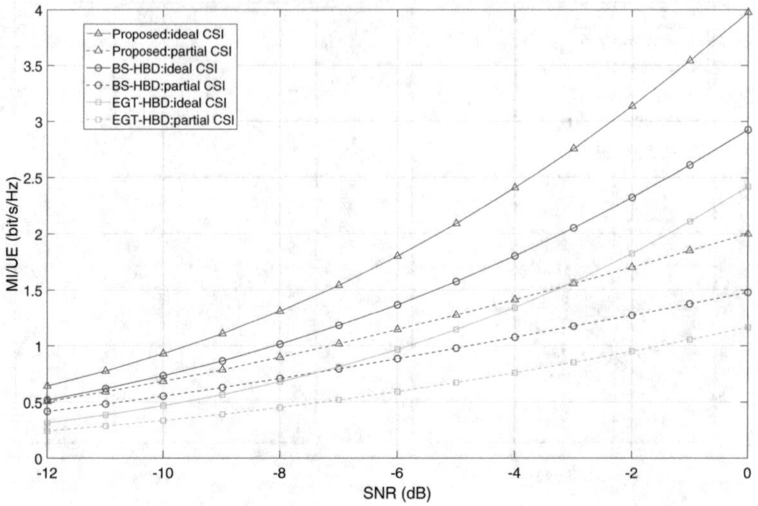

Fig. 4.8 MI comparison in CDL-based channel model

4.5.5 *MI in Other Configurations*

To verify a more general effectiveness of the proposed method, we adopt the 3GPP type-A clustered-delay-line (CDL) model, with only the azimuth angles considered in the ULA setup. More details about how to generate the type-A CDL-based channel can be found in Chapter 7 of [21]. With perfect CSI, we see that the proposed method still performs best the highest MI. This result is within expectation because the CDL channel belongs to the studied generic geometric model.

Considering that perfect CSI might be unattainable at the BS, a common practice is to feed back the best beams. To this end, we set the transmit SNR for channel estimation as 10 dB. Only the best one-third of beams will be reported together with their angles quantized into five bits. As can be seen from Fig. 4.8, the proposed HBD yields the highest MI with limited feedback. And the achieved MI is even well-comparable to other HBD candidates with ideal CSI at low-to-medium SNR.

4.6 Discussions and Summary

In this chapter, a systematic transceiver solution has been tailored for wMU mmWave mMIMO based on the criterion of MI maximization. The proposed scheme follows the popular HBD-based philosophy but avoids the potential ad-hoc or empirical nature shared by nearly all existing works along this line. For the limit case, we have demonstrated the asymptotic optimality of HBD in the sense of MI for wMU mmWave MIMO transceiver design. For the general case, we have devised high-performance HBD-wMU transceivers for both the MAS and the MBS, regardless of the resolution of the analog beamformers or the sparsity of the mmWave channels.

Since the investigated scenario is restricted to a centralized wMU mmWave system, future work will consider the network de-centralization. Regarding this topic, some promising directions include, but are not limited to, the advanced deep learning based MIMO transmission and cell-free mMIMO optimization.[2]

References

1. T.S. Rappaport, J. N. Murdock, F. Gutierrez, State of the art in 60-GHz integrated circuits and systems for wireless communications. Proc. IEEE. **99**(8), 1390–1436 (2011)
2. F. Rusek, D. Persson, B.K. Lau, E. Larsson, T. Marzetta, O. Edfors, F. Tufvesson, Scaling up MIMO: Opportunities and challenges with very large arrays. IEEE Signal Process. Mag. **30**(1), 40–60 (2013)
3. F. Boccardi, R. Heath, A. Lozano, T.L. Marzetta, P. Popovski, Five disruptive technology directions for 5G. IEEE Commun. Mag. **52**(2), 74–80 (2014)
4. J. Fernandez, N. Prelcic, K. Venugopal, R. Heath, Frequency-domain compressive channel estimation for frequency-selective hybrid mmWave MIMO systems. IEEE Trans. Wireless Commun. **17**(5), 2946–2960 (2018)
5. S. Gao, X. Cheng, L. Yang, Estimating doubly-selective channels for hybrid mmwave massive MIMO systems: a doubly-sparse approach. IEEE Trans. Wirel. Commun. **19**(9), 5703–5715 (2020)
6. O.E. Ayach, S. Rajagopal, S. Abu-Surra, Z. Pi, R. Heath, Spatially sparse precoding in millimeter wave MIMO systems. IEEE Trans. Wirel. Commun. **13**(3), 1499–1513 (2014)
7. X. Gao, L. Dai, S. Han, I. Chih-Lin, R. Heath, Energy-efficient hybrid analog and digital precoding for mmwave MIMO systems with large antenna arrays. IEEE J. Sel. Areas Commun. **34**(4), 998–1099 (2016)
8. S. Gao, X. Cheng, L. Yang, Spatial multiplexing with limited RF chains: Generalized beamspace modulation (GBM) for mmwave massive MIMO. IEEE J. Sel. Areas Commun. **37**(9), 2029–2039 (2019)
9. A. Alkhateeb, R. Heath, Frequency-selective hybrid precoding for limited feedback millimeter wave systems. IEEE Trans. Commun. **64**(5), 1801–1818 (2016)
10. A. Alkhateeb, G. Leus, R. Heath, Limited feedback hybrid precoding for multi-user millimeter wave systems. IEEE Trans. Wirel. Commun. **14**(11), 6481–6494 (2015)
11. W. Ni, X. Dong, Hybrid block diagonalization for massive multi-user MIMO systems. IEEE Trans. Commun. **64**(1), 201–211 (2016)
12. S. Gao, X. Cheng, L. Yang, Hybrid multi-user precoding for mmwave massive MIMO in frequency-selective channels, in *Proceeding of IEEE Wireless Communication and Networking Conference (WCNC), Virtual* (2020)
13. L. Liang, W. Xu, X. Dong, Low-complexity hybrid precoding in massive multi-user MIMO systems. IEEE Wirel. Commun. Lett. **35**(5), 653–656 (2014)
14. R. Rajashekar, L. Hanzo, Iterative matrix decomposition aided block diagonalization for mmWave multiuser MIMO systems. IEEE Trans. Wireless Commun. **16**(3), 1372–1384 (2017)
15. H. Yuan, J. An, N. Yang, K. Yang, T. Duong, Low-complexity hybrid precoding for multi-user millimeter wave systems over frequency-selective channel. IEEE Trans. Veh. Technol. **68**(1), 983–987 (2019)

[2] The readers are suggested to refer to [22] for further details about cell-free MIMO.

16. K. Venugopal, A. Alkhateeb, N. Gonzalez-Prelcic, R. Heath, Channel estimation for hybrid architecture-based wideband millimeter wave systems. IEEE J. Sel. Areas Commun. **35**(9), 1996–2009 (2017)
17. A. Sayeed, J. Brady, Beamspace MIMO for high-dimensional multiuser communication at millimeter-wave frequencies, in *Proceeding of Global Telecommunications Conference (Globecom), Atlanta, GA* (2013)
18. Z. Wang, M. Li, Q. Liu, A. Swindlehurs, Hybrid precoder and combiner design with low-resolution phase shifters in mmWave MIMO systems. IEEE J. Sel. Topics Signal Process. **12**(2), 256–269 (2018)
19. Q.H. Spencer, A.L. Swindlehurst, M. Haardt, Zero-forcing methods for downlink spatial multiplexing in multi-user MIMO channels. IEEE Trans. Signal Process. **52**(2), 461–471 (2004)
20. Q. Zhang, Y. Liu, G. Xie, J. Gao, K. Liu, An efficient hybrid diagonalization for multi-user mmWave massive MIMO systems, in *Proceeding of Global Symposium on Millimeter Waves (GSMM), Boulder, CO* (2018)
21. 3GPP, TR 38.900 study on channel model for frequency spectrum above 6 GHz, version 14.2.0 Release 14, 2017. Available Online: https://www.3gpp.org/DynaReport/38-series.htm
22. M. Bashar, K. Cumanan, A. Burr, H. Ngo, L. Hanzo, P. Xiao, On the performance of cell-free massive MIMO relying on adaptive NOMA/OMA mode-switching. IEEE Trans. Commun. **68**(2), 792–810 (2020)

Chapter 5
Millimeter-Wave Index Modulation for Vehicular Uplink Access

Abstract This chapter works on proposing an advanced index modulation (IM) scheme to underpin mmWave vehicular uplink access. The designed IM roots from a spectrum efficiency (SE-) enhanced technique, namely generalized beamspace modulation (GBM), but has generalized it from the narrow-band to realistic wideband scenarios via a symbol-based modulating framework. Besides, the result wideband GBM (wGBM) is accompanied by a dedicated Doppler compensation module to enhance its robustness against Doppler. The proposed single-user wGBM is further extended to a wideband multi-user (wMU) scenario to support massive connection and ultra-fast speed for the next-generation cellular. A well-design detector is carefully devised for in wMU-wGBM based on the message-passing algorithm. It can effectively alleviate the computational complexity and maintain a near-optimal performance in large-scale systems. Theoretical analyses and numerical simulations have been carried out to validate the advantages of wGBM in terms of error performance and energy efficiency.

Keywords mmWave · Massive multiple-input multiple-output (mMIMO) · Index modulation · Spectrum efficiency · Wideband generalized beamspace modulation · Message-passing algorithm

5.1 Introduction of Index Modulation (IM)

Thanks to the enhanced performance in terms of the bit error rate (BER) and power efficiency, index modulation has been receiving increasing research interests in recent years [1]. The main idea of IM is to utilize the activation status of antennas or subcarriers to convey the so-called index information. Up to date, there are two typical IM schemes, namely spatial modulation (SM) [2, 3] and subcarrier IM [4]. Based on these two prototypes, some other variants such as GSM [5], DSM [6], PDSM [7], DSFM [8], have been proposed for different application scenarios. All these techniques have demonstrated their advantages in the current centimeter-wave (cmWave) communication systems.

X. Cheng et al., *mmWave Massive MIMO Vehicular Communications*,
Wireless Networks, https://doi.org/10.1007/978-3-030-97508-1_5

Nowadays, mmWave massive multiple-input multiple-output (mMIMO) has been recognized as a key technique to provide Gbps data rate for next-generation wireless communications [9–11]. But unlike sub-6GHz digital MIMO, mmWave mMIMO transceivers adopt a hybrid structure to reduce the power consumption and hardware cost [12, 13]. Meanwhile, different from the typically isotropic environment by using cmWave as the wireless medium, mmWave channels are well known to exhibit limited scattering, thus the highly correlated channels may severely affect the error performance [14]. As a result, a simple migration of existing cmWave IM techniques into mmWave mMIMO is not feasible. Instead, the ultimate solution requires a judicious design by accounting for the unique properties of mmWave mMIMO. In mmWave mMIMO, the potential multiplexing gain (MG) is fundamentally limited by the minimum number of radio frequency (RF) chains at both ends. To cope with this fundamental limit and further boost the spectral efficiency (SE), there is an urgent need to develop IM techniques suitable for mmWave mMIMO. We will first list existing options, and then introduce our approach.

5.1.1 IM in Spatial-Domain

A natural implementing space is the spatial domain; that is, the index bits directly determine which antennas are activated. In [15], an antenna-group (AG-) GSM is designed for mmWave MIMO, with transmitter structure shown in Fig. 5.1. Clearly, in mMIMO, directly (de-)activating each and every antenna will incur unbearable complexity, together with a huge number of RF chains. AG-GSM is adapted to hybrid mmWave structures by (de-)activating groups of (as opposed to individual) antennas. However, this approach essentially divides the entire array to a few groups of smaller ones, and will thus suffer from a severe loss of array gain and angle resolution. In addition, as the MG is dictated by the number of groups, there is clearly a trade-off between the achievable MG and the array gain/angle resolution.

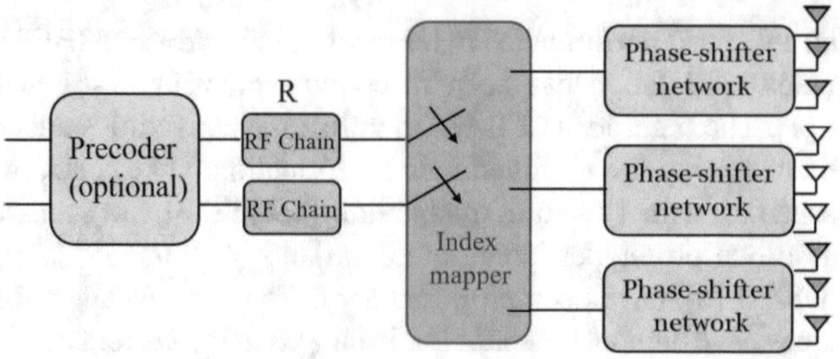

Fig. 5.1 The transmitter of mmWave AG-GSM

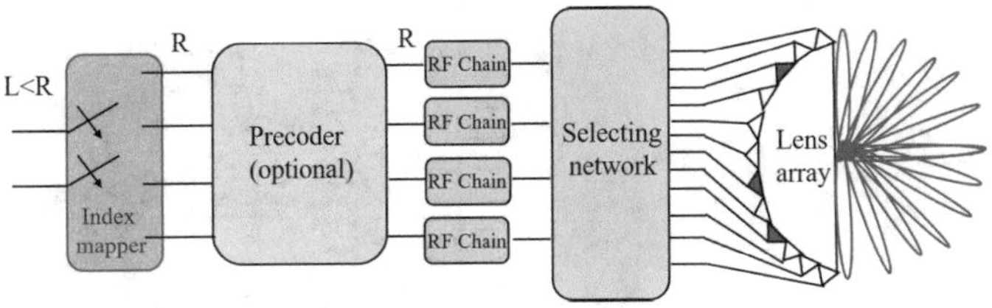

Fig. 5.2 The transmitter of mmWave EDC-IM

5.1.2 *IM in Digital-Domain*

Recall that GSM is essentially a digital technique, thus can be directly applied to the equivalent digitized channel (EDC) that is encountered before the RF chains at the transmitter (Tx) and after the RF chains at the receiver (Rx). The Tx structure of EDC-IM is shown in Fig. 5.2. Although only a subset of RF chains are activated when performing IM, all antennas are employed for transmission. Hence EDC-IM is not only applicable to hybrid structures, but can also fully exploit the large array gain. However, its maximum MG is clearly limited by the minimum number of RF chains at the transceivers.

5.1.3 *IM in Beamspace-Domain*

In view of the limitations of the abovementioned options, one may have realized that a proper domain has to leverage both the channel properties and the hybrid hardware structures that are unique to mmWave mMIMO [16–18]. Reference [16] proposed a spatial scattering modulation (SSM) for uplink mmWave communications and studied its BER performance. Adopting analog beamforming, [17] proposed a low-complexity beam index modulation (BIM) for mmWave MIMO. chains. It should be noted that the SSM in [16] and BIM in [17] only consider the EBC with single input, and the optimization of EBC has not been touched. Instead, the proposed GBM system in [18] shown in Fig. 5.3 that gives a generalized case without any constraints on the activated RF chains, and the EBC is further extended to the MIMO scenario. Specially, the index mapping of GBM occurs after the RF chains but before the selecting network. Different from both aforementioned options, all RF chains and all antennas at the Tx are always active. As a result, not only that the array gain is fully exploited, but also the achievable MG is no longer restricted by the number of RF chains. The resultant GBM design is also perfectly compatible with prevalent mmWave mMIMO systems.

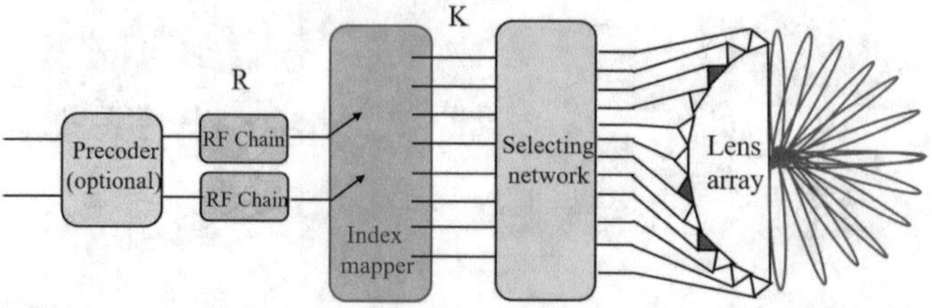

Fig. 5.3 The transmitter of mmWave GBM

5.2 Wideband Generalized Beamspace Modulation (wGBM)

5.2.1 Design Motivation

Although GBM demonstrates remarkable superiority, its applicability is somewhat limited because it was designed for narrowband channels, while the practical mmWave channels are generally wideband exhibiting strong frequency selectivity [19–21]. Many works have revealed that with the help of OFDM, extending IM to wideband channels in conventional fully-digital transceivers is straightforward by applying the corresponding narrowband IM on each subcarrier independently [7, 8]. However, the mentioned similar approach fails for the broadband expansion of GBM due to two main reasons: (1) the analog precoder in hybrid OFDM systems is shared by all subcarriers, therefore extending GBM requires joint consideration among all subcarriers; and (2) even with a decent analog precoder, simply inheriting GBM's index mapper will render the SE-enhanced merit disappearing in OFDM systems.

In light of the aforementioned challenges, we innovatively propose a symbol-based method to guide the subsequent wGBM design [22], such that the compatibility with hybrid OFDM systems can be guaranteed without losing the SE-enhanced advantage. Within the symbol-based framework, we position an FFT module before the OFDM modulator and an RF-chain-reduced digital-to-analog module after the OFDM modulator. As a result, wGBM does not occur on individual subcarriers but across all subcarriers, or equivalently, an entire OFDM symbol. This strategy makes it feasible to use fewer RF chains to gain a higher SE as GBM does. Due to the symbol-based modulation, wGBM potentially entails significant detection complexity. For implementation considerations, a simple linear minimum mean square error (LMMSE)-based detector is accordingly devised by transforming the complicated symbol (block) detection into the more manageable subcarrier (sub-block) detection. To optimize the error performance under the LMMSE-based detector, a low-complexity yet near-optimal beam selection method is proposed by leveraging the channel's sparsity in beamspace. Finally, to accommodate time-varying channels, we carefully position a first-order compensator at the receiver that can significantly combat Doppler at a negligible cost.

To the best of our knowledge, there is no IM technique designed for hybrid wideband mmWave systems except for the hybrid precoded spatial modulation (hPSM) proposed in [23]. Although hPSM is able to improve the error performance, its SE is still restricted by the number of RF chains due to its subcarrier-based design methodology. In this sense, the proposed wGBM is the first SE-enhanced wideband IM for hybrid mmWave systems and can be well-suited for time-varying channels. Therefore, wGBM is a promising uplink transmission candidate for mobile mmWave systems. Next, we will introduce the wGBM mmWave system and channel models, wGBM transceiver design, the Doppler compensator for wGBM, simulation results and summary in detail.

5.2.2 System and Channel Models

In this section, we will introduce the system and channel models of the studied wideband mmWave mMIMO, based on which input-output (I-O) relationship, we will present the system.

We consider an uplink wideband mmWave mMIMO system, where N_m-dimensional lens-array antennas with M_m ($M_m << N_m$) RF chains and N_b-dimensional lens-array antennas with M_b ($M_b << N_b$) RF chains are deployed at the transmitter (mobile station) and receiver (base station), respectively. In the studied system shown in Fig. 5.4, the mobile station communicates with BS via L ($M_m \leq L \leq N_m$) streams. Without loss of generality, we further assume $L = M_b$ and adopt OFDM for wideband transmission, with its number of subcarriers being K and the length of cyclic prefix (CP) being L_{cp}. All the channel and signal related notations are listed in Table 5.1 for readers' reference.

Let $\boldsymbol{s}_k$ be the transmitted signal on subcarrier k. Before transmission, an inverse FFT (IFFT) operation is first applied to $\boldsymbol{s}_k$, followed by CP insertion. The processed signal then goes through the devised digital-to-analog module and finally be precoded by the analog precoder

$$\boldsymbol{P}_A = [\boldsymbol{p}_T(t_1), \boldsymbol{p}_T(t_2), \cdots, \boldsymbol{p}_T(t_L)], \tag{5.1}$$

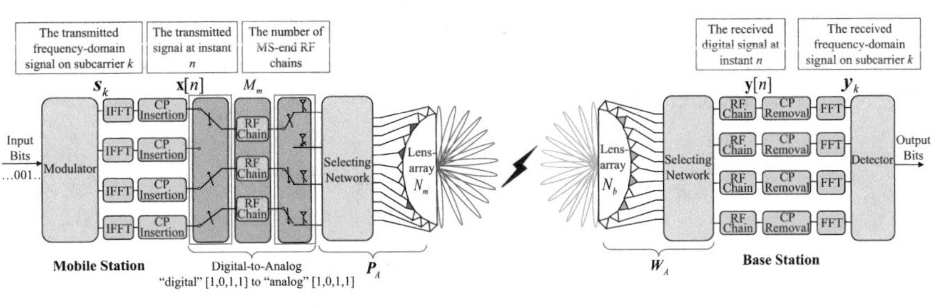

Fig. 5.4 Illustration of uplink wideband mmWave mMIMO system

Table 5.1 Notations relating to uplink mmWave mMIMO system

Notation			
Type	Time domain	Frequency domain	Definition
Channel	$\mathbf{H}_d[n]$	$\boldsymbol{H}_k[r]$	Spatial channel ($\mathbb{C}^{N_b \times N_m}$, $w_p \neq 0$)
	$\mathbf{H}_d$	$\boldsymbol{H}_k$	Spatial channel ($\mathbb{C}^{N_b \times N_m}$, $w_p = 0$)
	$\bar{\mathbf{H}}_d[n]$	$\bar{\boldsymbol{H}}_k[r]$	Equivalent channel incorporating analog parts ($\mathbb{C}^{L \times L}$, $w_p \neq 0$)
	$\bar{\mathbf{H}}_d$	$\bar{\boldsymbol{H}}_k$	Equivalent channel incorporating analog parts ($\mathbb{C}^{L \times L}$, $w_p = 0$)
Signal	$\mathbf{x}[n]$	$\boldsymbol{s}_k$	Signal before analog precoding
	$\mathbf{y}[n]$	$\boldsymbol{y}_k$	Signal after analog combing
	$\bar{\mathbf{y}}[n]$	$\bar{\boldsymbol{y}}_k$	Signal after compensation

where $\boldsymbol{p}_T(i)$ stands for the i-th column of $\boldsymbol{F}_{N_m}$. It is worth mentioning that $\boldsymbol{P}_A$ remains identical across all subcarriers in wideband hybrid system, forcing the subsequent wGBM design to be designed across all subcarriers. As will be detailed later, it is such a unique feature that renders the proposed wGBM distinctive from the existing wideband IMs.

To incorporate the angular-delay sparsity as well as the frequency-time selectivity exhibited by mmWave channels, we adopt a generic geometric channel model with N_c distinguishable delay taps and P dominant paths. According to [24], the tap-d channel at time instant-n $\mathbf{H}_d[n] \in \mathcal{C}^{N_b \times N_m}$ can be written as

$$\mathbf{H}_d[n] = \sqrt{\frac{N_b N_m}{P}} \sum_{p=1}^{P} \alpha_p p_{rc}(dT_s - \tau_p) \boldsymbol{a}_r(\theta_p) \boldsymbol{a}_t^*(\phi_p) e^{j w_p n}, \tag{5.2}$$

where T_s is the sampling period; $p_{rc}(\cdot)$ is the response of the pulse-shaper. For the p-th path, τ_p is the propagation delay uniformly distributed within $[0, (N_c - 1)T_s)$; $w_p = 2\pi f_c v_m T_s \sin(\theta_p)/c_v$ is its associated Doppler frequency shift, with f_c being the system carrier frequency, c_v being the speed of light, and v_m being the relative speed between mobile station (MS) and base station (BS); $\alpha_p \sim \mathcal{CN}(0, 1)$ is the complex gain; θ_p and ϕ_p represent the angle of arrival (AoA) and angle of departure (AoD), respectively. $\boldsymbol{a}_t(\cdot) \in \mathbb{C}^{N_m \times 1}$ and $\boldsymbol{a}_r(\cdot) \in \mathbb{C}^{N_b \times 1}$ stand for the MS- and BS-end array responses, respectively. Define

$$\boldsymbol{f}_N(\psi) = \frac{1}{\sqrt{N}} \left[1, e^{j\psi}, \cdots, e^{j(N-1)\psi}\right]^T. \tag{5.3}$$

Then, under the commonly used half-wavelength spaced uniform linear arrays (ULAs), we have $\boldsymbol{a}_t(\phi_p) = \boldsymbol{f}_{N_m}(\pi \sin(\phi_p))$ and $\boldsymbol{a}_r(\theta_p) = \boldsymbol{f}_{N_b}(\pi \sin(\theta_p))$. In this section, channel state information (CSI) is assumed to be available at the transceivers. This assumption is reasonable, because CSI can be accurately estimated in doubly-selective mmWave mMIMO channels by exploiting the delay-angular sparsity [24].

Given the time-domain channel response in Eq. (5.2), the corresponding inter-subcarrier channel response $\boldsymbol{H}_k[r] \in \mathcal{C}^{N_b \times N_m}\,(k, r \in [0, K-1])$ can be expressed as[25]

$$\boldsymbol{H}_k[r]=\frac{1}{K}\sum_{i=0}^{K-1}\sum_{d=0}^{N_c-1}\mathbf{H}_d[L_{cp}+i]e^{-j2\pi(rd+(k-r)i)/K}. \tag{5.4}$$

where $\boldsymbol{H}_k[r]$ represents the channel response at subcarrier-k if $k = r$, or the inter-carrier interference (ICI) from subcarrier-r to k if $k \neq r$.

Apparently, when $w_p = 0$, $\mathbf{H}_d[n]$ turns out to be time-invariant. Denote the static version of $\mathbf{H}_d[n]$ as $\mathbf{H}_d$, then in this case, all the inter-subcarrier channels vanish except for leaving

$$\boldsymbol{H}_k[k]= \boldsymbol{H}_k = \sum_{d=0}^{N_c-1}\mathbf{H}_d e^{-j\frac{2\pi k}{K}d}. \tag{5.5}$$

After channel propagation, the received signal on subcarrier k is given by

$$\boldsymbol{r}_k = \boldsymbol{H}_k[k]\boldsymbol{P}_A\boldsymbol{s}_k + \sum_{r=0,r\neq k}^{K-1}\boldsymbol{H}_k[r]\boldsymbol{P}_A\boldsymbol{s}_r + \boldsymbol{w}_k, \tag{5.6}$$

where $\boldsymbol{w}_k \sim \mathcal{CN}\left(\mathbf{0}, \sigma^2\boldsymbol{I}_{N_b}\right)$ is the white Gaussian noise, with σ^2 representing the noise power. $\boldsymbol{r}_k$ is first combined by the analog combiner

$$\boldsymbol{W}_A = [\boldsymbol{f}_R(r_1), \boldsymbol{f}_R(r_2), \cdots, \boldsymbol{f}_R(r_L)] \in \mathcal{C}^{N_b \times L},$$

with $\boldsymbol{f}_R(i)$ being the i-th column of $\boldsymbol{F}_{N_b}$. After CP removal and FFT operation, the signal to be processed in digital baseband becomes

$$\begin{aligned}\boldsymbol{y}_k &= \boldsymbol{W}_A^*\boldsymbol{H}_k[k]\boldsymbol{P}_A\boldsymbol{s}_k + \sum_{r=0,r\neq k}^{K-1}\boldsymbol{W}_A^*\boldsymbol{H}_k[r]\boldsymbol{P}_A\boldsymbol{s}_r + \boldsymbol{W}_A^*\boldsymbol{w}_k \\ &= \bar{\boldsymbol{H}}_k[k]\boldsymbol{s}_k + \sum_{r=0,r\neq k}^{K-1}\bar{\boldsymbol{H}}_k[r]\boldsymbol{s}_r + \boldsymbol{\xi}_k,\end{aligned} \tag{5.7}$$

where $\boldsymbol{\xi}_k = \boldsymbol{W}_A^*\boldsymbol{w}_k \sim \mathcal{CN}\left(\mathbf{0}, \sigma^2\boldsymbol{I}_L\right)$ is the combined noise. Finally, the system I-O relationship can be expressed as

$$\underbrace{\begin{bmatrix} y_0 \\ \vdots \\ y_{K-1} \end{bmatrix}}_{y} = \underbrace{\begin{bmatrix} \bar{H}_0[0] & \cdots & \bar{H}_0[K-1] \\ \vdots & \ddots & \vdots \\ \bar{H}_{K-1}[0] & \cdots & \bar{H}_{K-1}[K-1] \end{bmatrix}}_{H} \underbrace{\begin{bmatrix} s_0 \\ \vdots \\ s_{K-1} \end{bmatrix}}_{s} + \underbrace{\begin{bmatrix} \xi_0 \\ \vdots \\ \xi_{K-1} \end{bmatrix}}_{\xi}. \tag{5.8}$$

Define $\bar{H}_k = W_A^* H_k P_A$. When $w_p = 0$, H in Eq. (5.8) can be further simplified as

$$\bar{H} = \mathbf{diag}\{\bar{H}_k\}_{k=0}^{K-1}. \tag{5.9}$$

5.2.3 wGBM Transceiver Over Static Channels

In this section, we first detail the general wGBM transceiver design in the absence of Doppler. Then we will carry out theoretical analyses for wGBM in terms of the error performance and EE.

The prototype of wGBM modulator shown in Fig. 5.5 comprises bit splitter, bit mapper, FFT module, OFDM modulator and digital-to-analog module. To be specific, the input bit stream is split into K branches, each having $b = M_m \log_2(M) + \left\lfloor \log_2(\mathbb{C}_L^{M_m}) \right\rfloor$ bits. Taking the k-th branch as an example to explain the specifics. The first $M_m \log_2(M)$ bits will be modulated into a symbol vector $u_k = [u_{k,1}, u_{k,2}, \cdots, u_{k,M_m}]^T$, of which each entry is chosen from a normalized M-ary phase shift keying/quadrature amplitude (PSK/QAM) constellation χ. The remaining $\left\lfloor \log_2(\mathbb{C}_L^{M_m}) \right\rfloor$ bits will be mapped into a length-M_m sequence $I_k \subset S = \{S_1, \cdots, S_Q\}$, whose elements range from $[1, L]$ in an ascending order. As a result, the data vector after bit mapper can be written as

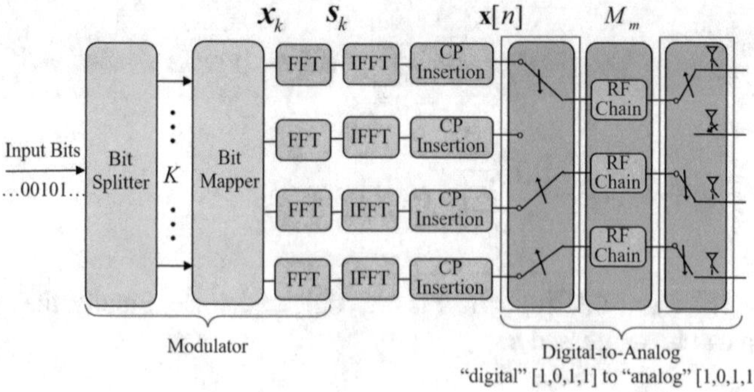

Fig. 5.5 Illustration of wGBM's modulator

$$\boldsymbol{x}_k = \boldsymbol{I}_L[:, I_k]\boldsymbol{u}_k. \tag{5.10}$$

After FFT module, a stacked frequency-domain signal vector is obtained as

$$\begin{aligned}\boldsymbol{s} &= [\boldsymbol{s}_0^T, \boldsymbol{s}_1^T, \cdots, \boldsymbol{s}_{K-1}^T]^T \\ &= (\boldsymbol{F}_K \otimes \boldsymbol{I}_L)\boldsymbol{x},\end{aligned} \tag{5.11}$$

with $\boldsymbol{x}$ being a concatenation of $\boldsymbol{x}_k$. By applying IFFT, the corresponding time-domain signal becomes

$$\boldsymbol{x} = (\boldsymbol{F}_K \otimes \boldsymbol{I}_L)^* \boldsymbol{s}. \tag{5.12}$$

Define $\widehat{x}$ to be the concatenated signal after CP insertion. Then the corresponding time-domain signal at instant n is

$$\mathbf{x}[n] = \widehat{x}[nL + 1 : (n+1)L](n < K + L_{cp} - 1). \tag{5.13}$$

Note that, with FFT applied before the OFDM modulator, $\mathbf{x}[n]$ always contains a fixed number of zero elements. Therefore, we can employ a simple digital-to-analog module to reduce the use of RF chains. That is, only those non-zero elements of $\mathbf{x}[n]$ will be up-converted via RF chains while the zero elements will be connected to the analog ground (essentially transmitting 0). It is worth mentioning that wGBM degenerates to GBM when $K = 1$. By taking the CP consumption into account, the achievable SE of wGBM can be expressed as

$$\eta = \frac{K}{K + L_{cp}} \left(\left\lfloor \log_2 \mathbb{C}_L^R \right\rfloor + R\log_2 M \right). \tag{5.14}$$

(1) ML detector: Based on Eqs. (5.8) and (5.11), the ML detector can be expressed as

$$\begin{aligned}\hat{\boldsymbol{x}} &= \underset{\forall k, \boldsymbol{x}_k \in G}{\arg\min} \left\| \boldsymbol{y} - \bar{\boldsymbol{H}}\boldsymbol{s} \right\|^2 \\ &= \underset{\forall k, \boldsymbol{x}_k \in G}{\arg\min} \left\| \boldsymbol{y} - \bar{\boldsymbol{H}}(\boldsymbol{F}_K \otimes \boldsymbol{I}_L)\boldsymbol{x} \right\|^2,\end{aligned} \tag{5.15}$$

where G contains all possible $\boldsymbol{x}_k$'s. Since the detection is performed across all subcarrier symbols, the computational complexity in terms of multiplications is $O \sim (2^{bK}(4K^2R + 4KL^2 + 2KL))$, which grows exponentially with K. The complexity would be badly high, driving us to seek a low-complexity alternative for practice use.

(2) LMMSE-based detector: Recall that the high complexity of ML detector is resulted from the symbol-based nature of wGBM. Therefore, to reduce the complexity, one has to decouple the intertwined relationship among all subcarriers.

Towards this objective, a two-step low-complexity detector is proposed. First, we apply the widely-used LMMSE equalizer $\boldsymbol{M} = \{\bar{\boldsymbol{H}}^* \bar{\boldsymbol{H}} + \sigma^2 \boldsymbol{I}_{LK}\}^{-1} \bar{\boldsymbol{H}}^*$ to get an estimated $\boldsymbol{s}$, i.e., $\bar{\boldsymbol{s}} = \boldsymbol{M}\boldsymbol{y}$. Leveraging Eq. (5.11), $\boldsymbol{x}_k$ can accordingly be estimated as

$$\begin{aligned} \bar{\boldsymbol{x}}_k &= (\boldsymbol{F}_K{}^*[k, :] \otimes \boldsymbol{I}_L)\bar{\boldsymbol{s}} \\ &= (\boldsymbol{F}_K{}^*[k, :] \otimes \boldsymbol{I}_L)\boldsymbol{M}\boldsymbol{y}. \end{aligned} \tag{5.16}$$

Next, the ML detection is applied to the low-dimensional vector $\bar{\boldsymbol{x}}_k$, giving rise to the demodulated data vector

$$\hat{\boldsymbol{x}}_k = \arg\min_{\boldsymbol{x}_k \in G} \|\bar{\boldsymbol{x}}_k - \boldsymbol{x}_k\|^2. \tag{5.17}$$

The detection complexity in terms of multiplications is $O \sim (2KL^3 + 8K^2L^3 + K(4KL + 2^{b+1}L))$, which is a polynomial one with K and L. In essence, the proposed LMMSE-based detector transforms the complicated symbol (block) detection into the more manageable subcarrier (sub-block) detection, thereby being more suitable for practical implementation.

(3) AMP-based detector: LMMSE exhibits a cubic-order complexity with respect to the signal dimensionality. To further lower the detection burden, the AMP algorithm comes into play to help reduce the complexity to square order. In this chapter, we follow a similar implementing framework proposed in [26]. Specifically, the AMP-based detector aims to decouple a complicated KL-dimension joint estimation problem into KL simple ones, each dealing with a scalar estimation. The detailed procedures of the proposed AMP-based detector is summarized as Algorithm 3.

Recall that those L beams for constructing the analog parts should be determined in advance. Based on the estimated frequency-domain signal vector $\bar{\boldsymbol{s}}$, we can obtain the beam selection criterion of minimizing the mean square error function

$$\|\boldsymbol{e}\|^2 = \|\bar{\boldsymbol{s}} - \boldsymbol{s}\|^2 = \|\boldsymbol{M}\boldsymbol{y} - \boldsymbol{s}\|^2. \tag{5.18}$$

Ideally, those L beams can be obtained via an exhaustive search within the set of all possible beam candidates, but the computational complexity would be tremendously high under mMIMO. Thanks to the fact that beamspace sparsity holds in wideband channels as well, we can apply the power-based criterion proposed in [18] on each subcarrier to get a small-size beam set P_k. Define $P = \bigcup_{\forall k} P_k$ and set $(\boldsymbol{r}, \boldsymbol{t}) = \{(r_1, t_1), \cdots, (r_L, t_L)\} \subset \mathcal{P}$ with $\forall i$, $(r_i, t_i) \in \mathcal{P}$, then the L beam indices for constructing the analog beamformer can be determined as

$$\{(\bar{r}_1, \bar{t}_1), \cdots, (\bar{r}_L, \bar{t}_L)\}$$

Algorithm 3 AMP-based detector for wGBM

1: **Initialization:** t=1, $\tilde{\boldsymbol{H}} = \bar{\boldsymbol{H}}(\boldsymbol{F}_K \otimes \boldsymbol{I}_L)$;
$\hat{x}_i^1 = \sum_{x\in\chi} x/2^b$, $\hat{v}_i^1 = \sum_{x\in\chi} \left|x - \hat{x}_i^1\right|^2/2^b$, $i = 1, \cdots, KL$; $V_j^0 = 1, Z_j^0 = \boldsymbol{y}(j), j = 1, \cdots, KL$;

2: **Decoupling step:** For $j = 1, \cdots, KL$ and $i = 1, \cdots, KL$, compute
$V_j^t = \delta \sum_i \left|\tilde{\boldsymbol{H}}[j,i]\right|^2 \hat{v}_i^t + (1-\delta)V_j^{t-1}$,
$Z_j^t = \sum_i \tilde{\boldsymbol{H}}[j,i]\hat{x}_i^t - V_j^t(\boldsymbol{y}(j) - Z_j^{t-1})/(\sigma^2 + V_j^{t-1})$
For $i = 1, \cdots, KL$, calculate
$\Sigma_i^t = \left[\sum_j \frac{\left|\tilde{\boldsymbol{H}}[j,i]\right|^2}{\sigma^2+V_j^t}\right]^{-1}$, $\bar{x}_i^t = \delta\hat{x}_i^t + (1-\delta)\bar{x}_i^{t-1}$
$R_i^t = \bar{x}_i^t + \Sigma_i^t \sum_j \tilde{\boldsymbol{H}}[j,i]^*(\boldsymbol{y}(j) - Z_j^t)/(\sigma^2 + V_j^t)$
Here $\delta = 1$ is used in the first iteration.

3: **Denoising step:** For $i = 1, \cdots, KL$, $\tilde{k} = ceil(i/L) - 1$, $\bar{S}_q = \{1, \cdots, L\}\backslash S_q$, $q = 1, \cdots, Q$

$$D_i^t(x_i) = \exp\left(-\frac{|x_i|^2 - 2\mathrm{Re}(R_i^t x_i^*)}{\Sigma_i^t}\right) \times \sum_{q:(i-\tilde{k}L)\in S_q} \prod_{\substack{a\in S_q,\\ a\neq(i-\tilde{k}L)}} \left(\sum_{x\in\chi} \exp\left(-\frac{|x|^2 - 2\mathrm{Re}(R_{\tilde{k}L+a}^t x^*)}{\Sigma_{\tilde{k}L+a}^t}\right)\right)$$

$$E_i^t = \sum_{q:(i-\tilde{k}L)\in\bar{S}_q} \prod_{a\in S_q} \left(\sum_{x\in\chi} \exp\left(-\frac{|x|^2 - 2\mathrm{Re}(R_{\tilde{k}L+a}^t x^*)}{\Sigma_{\tilde{k}L+a}^t}\right)\right)$$

$$q^t(x_i) = \begin{cases} \frac{D_i^t(x_i)}{\sum_{x\in\chi} D_i^t(x) + E_i^t}, & x_i \in \chi \\ \frac{E_i^t}{\sum_{x\in\chi} D_i^t(x) + E_i^t}, & x_i = 0 \end{cases}$$

$$\hat{x}_i^{t+1} = \sum_{x\in\chi} x q^t(x_i = x)$$

$$\hat{v}_i^{t+1} = \sum_{x\in\chi} |x|^2 q^t(x_i = x) - |\hat{x}_i^{t+1}|^2$$

4: Set $t \leftarrow t+1$ and proceed to step (2) until $t < T_{max}$ or $\sum_i \left|\hat{x}_i^{t+1} - \hat{x}_i^t\right|^2 < 0.1$.

$$\begin{aligned}
&= \underset{\forall(\boldsymbol{r},\boldsymbol{t})\subset\mathcal{P}}{\arg\min}\, \mathbb{E}\left\{\|\boldsymbol{M}\boldsymbol{y} - \boldsymbol{s}\|^2\right\} \\
&= \underset{\forall(\boldsymbol{r},\boldsymbol{t})\subset\mathcal{P}}{\arg\min} \left\{\left\|\boldsymbol{M}\bar{\boldsymbol{H}} - \boldsymbol{I}_{LK}\right\|^2 + \sigma^2\|\boldsymbol{M}\|^2\right\} \\
&\overset{(a)}{\simeq} \underset{\forall(\boldsymbol{r},\boldsymbol{t})\subset\mathcal{P}}{\arg\min} \left\{\left\|\bar{\boldsymbol{H}}^+\bar{\boldsymbol{H}} - \boldsymbol{I}_{LK}\right\|^2 + \sigma^2\left\|\bar{\boldsymbol{H}}^+\right\|^2\right\} \\
&= \underset{\forall(\boldsymbol{r},\boldsymbol{t})\subset\mathcal{P}}{\arg\min}\, Tr\left\{\{\bar{\boldsymbol{H}}^*\bar{\boldsymbol{H}}\}^{-1}\right\},
\end{aligned} \tag{5.19}$$

where (a) is for $\boldsymbol{M} \simeq \bar{\boldsymbol{H}}^{+}$ at high signal-to-noise ratio (SNR). Such an approximation is reasonable as IM is well-known to manifest its advantage at high SNR [1]. Based on the selected beams, the analog precoder and combiner are accordingly designed as

$$\boldsymbol{P}_A = [\boldsymbol{f}_T(\bar{t}_1), \boldsymbol{f}_T(\bar{t}_2), \cdots, \boldsymbol{f}_T(\bar{t}_L)], \tag{5.20a}$$

$$\boldsymbol{W}_A = [\boldsymbol{f}_R(\bar{r}_1), \boldsymbol{f}_R(\bar{r}_2), \cdots, \boldsymbol{f}_R(\bar{r}_L)]. \tag{5.20b}$$

5.2.4 Performance Analysis

(1) *Error Performance*

After accomplishing the wGBM transceiver design, we then analyze its error performance under the proposed LMMSE detection. Due to the similarity across all subcarriers, the asymptotic pairwise error probability (APEP) of wGBM can be approximated as

$$\begin{aligned} P_{\text{APEP}} &= \frac{1}{Kb2^b} \sum_{k=0}^{K-1} \sum_{\boldsymbol{x}_k} \sum_{\hat{\boldsymbol{x}}_k} e(\boldsymbol{x}_k, \hat{\boldsymbol{x}}_k) P\big(\boldsymbol{x}_k \rightarrow \hat{\boldsymbol{x}}_k\big) \\ &\simeq \frac{1}{b2^b} \sum_{\boldsymbol{x}_k} \sum_{\hat{\boldsymbol{x}}_k} e(\boldsymbol{x}_k, \hat{\boldsymbol{x}}_k) P\big(\boldsymbol{x}_k \rightarrow \hat{\boldsymbol{x}}_k\big), \end{aligned} \tag{5.21}$$

where $e(\boldsymbol{x}_k, \hat{\boldsymbol{x}}_k)$ stands for the number of error bits between $\boldsymbol{x}_k$ and $\hat{\boldsymbol{x}}_k$, and $P\big(\boldsymbol{x}_k \rightarrow \hat{\boldsymbol{x}}_k\big)$ stands for the pair-wise error probability.

Proposition 1 *For wideband mmWave systems equipped with $M_m (M_b)$ RF chains at the MS (BS) and communicating through channels consisting of P dominant beam paths, the pair-wise error probability at subcarrier k can be approximated as*

$$\begin{aligned} P\big(\boldsymbol{x}_k \rightarrow \hat{\boldsymbol{x}}_k\big) \approx & \frac{P\mathbb{C}_{P-1}^{\nu-1}}{12} \mathbb{B}\left(\frac{N_b N_m A}{4\sigma^2 P} \sum_{l=1}^{L} \left|\Delta x_{k,l}\right|^2 + P - \nu + 1, \nu\right) \\ + & \frac{P\mathbb{C}_{P-1}^{\nu-1}}{4} \mathbb{B}\left(\frac{N_b N_m A}{3\sigma^2 P} \sum_{l=1}^{L} \left|\Delta x_{k,l}\right|^2 + P - \nu + 1, \nu\right), \end{aligned} \tag{5.22}$$

with $A = \int_0^{(N_c-1)T_s} \big(\frac{|p_{rc}(\lfloor\tau\rfloor-\tau)|^2 + |p_{rc}(\lfloor\tau\rfloor-\tau+T_s)|^2}{2(N_c-1)T_s} + \frac{(|p_{rc}(\lfloor\tau\rfloor-\tau)| - |p_{rc}(\lfloor\tau\rfloor-\tau+T_s)|)^2}{2(N_c-1)T_s}\big) d\tau$, and $\Delta x_{k,l}$ is the difference between the l-th element of $\boldsymbol{x}_k$ and $\hat{\boldsymbol{x}}_k$. ν denotes the diversity gain ranging from 1 to P.

Proof Leveraging the fast-decaying nature of p_{rc}, we get

$$\begin{aligned}
\delta_l &= \frac{N_b N_m}{P}\left|\sum_{d=0}^{N_c-1} \alpha_l p_{rc}(dT_s - \tau_l) e^{-jd\frac{2\pi k}{K}}\right|^2 \\
&\approx \frac{N_b N_m}{P} |\alpha_l p_{rc} \left(\lfloor \tau_l / T_s \rfloor T_s - \tau_l\right) e^{-j\lfloor \tau_l/T_s \rfloor \frac{2\pi k}{K}} \\
&\quad + \alpha_l p_{rc} \left(\left(\lfloor \tau_l / T_s \rfloor + 1\right) T_s - \tau_l\right) e^{-j(\lfloor \tau_l/T_s \rfloor + 1)\frac{2\pi k}{K}}|^2 \\
&\approx \frac{N_b N_m}{P} |\alpha_l|^2 \{(|p_{rc}\left(\lfloor \tau_l \rfloor - \tau_l\right)|^2 + |p_{rc}(\lfloor \tau_l \rfloor - \tau_l + T_s)|^2)/2 + \\
&\quad (|p_{rc}\left(\lfloor \tau_l \rfloor - \tau_l\right)| - |p_{rc}\left(\lfloor \tau_l \rfloor - \tau_l + T_s\right)|)^2/2\}. \qquad (5.23)
\end{aligned}$$

Define $\boldsymbol{R} = \mathbf{diag}\left\{\frac{\delta_l}{\delta_l + \sigma^2}\right\}_{l=1}^{L}$. According to Eq. (5.16), we have

$$\begin{aligned}
\bar{\boldsymbol{x}}_k &= (\boldsymbol{F}_K{}^*[k,:] \otimes \boldsymbol{I}_L)\boldsymbol{M}\bar{\boldsymbol{H}}\boldsymbol{s} + (\boldsymbol{F}_K{}^*[k,:] \otimes \boldsymbol{I}_L)\boldsymbol{M}\boldsymbol{\xi} \\
&= (\boldsymbol{F}_K{}^*[k,:] \otimes \boldsymbol{I}_L)(\boldsymbol{I}_K \otimes \boldsymbol{R})\boldsymbol{s} + \boldsymbol{n}_k \\
&= (\boldsymbol{I}_1 \otimes \boldsymbol{R})(\boldsymbol{F}_K{}^*[k,:] \otimes \boldsymbol{I}_L)\boldsymbol{s} + \boldsymbol{n}_k \\
&= \boldsymbol{R}\boldsymbol{x}_k + \boldsymbol{n}_k. \qquad (5.24)
\end{aligned}$$

Thus, the LMMSE-based detector in Eq. (5.17) can be approximated as

$$\hat{\boldsymbol{x}}_k \simeq \underset{\boldsymbol{x}_k \in G}{\arg\min} \|\bar{\boldsymbol{x}}_k - \boldsymbol{R}\boldsymbol{x}_k\|^2. \qquad (5.25)$$

The above equation can further lead to

$$\begin{aligned}
P\left(\boldsymbol{s}_k \to \hat{\boldsymbol{s}}_k\right) &= \mathbb{E}_{\bar{\boldsymbol{H}}_k}\left\{Q\left(\sqrt{\left(\left\|\boldsymbol{R}(\boldsymbol{x}_k - \hat{\boldsymbol{x}}_k)\right\|^2\right)^2/\sigma_k^2}\right)\right\} \\
&\overset{(b)}{\simeq} \mathbb{E}_{\bar{\boldsymbol{H}}_k}\left\{\frac{1}{12}\exp\left(-\left(\left\|\boldsymbol{R}(\boldsymbol{x}_k - \hat{\boldsymbol{x}}_k)\right\|^2\right)^2/2\sigma_k^2\right)\right\} \\
&\quad + \mathbb{E}_{\bar{\boldsymbol{H}}_k}\left\{\frac{1}{4}\exp\left(-2\left(\left\|\boldsymbol{R}\left(\boldsymbol{x}_k - \hat{\boldsymbol{x}}_k\right)\right\|^2\right)^2/3\sigma_k^2\right)\right\}, \qquad (5.26)
\end{aligned}$$

where (b) comes from $Q(x) \simeq 1/12e^{-x^2/2} + 1/4e^{-2x^2/3}$ [4], and $\sigma_k^2 = 2\sigma^2\sum_{l=1}^{L}\frac{\delta_l^3}{(\delta_l+\sigma^2)^4}\left|\Delta x_{k,l}\right|^2$. Let $\boldsymbol{\gamma} = [\gamma_1, \cdots, \gamma_L]^T$ with $\gamma_l = |\alpha_l|^2$, and $\boldsymbol{\beta} = [\beta_1, \cdots, \beta_L]^T$ with

$$\beta_l = (|p_{rc}\left(\lfloor \tau_l \rfloor - \tau_l\right)|^2 + |p_{rc}\left(\lfloor \tau_l \rfloor - \tau_l + T_s\right)|^2)/2 +$$

$$(|p_{rc}(\lfloor \tau_l \rfloor - \tau_l)| - |p_{rc}(\lfloor \tau_l \rfloor - \tau_l + T_s)|)^2/2. \tag{5.27}$$

Recall that γ_l obeys unit exponential distribution, so by replacing all γ_l's with the p_0-th ($1 \leq p_0 \leq P$) largest one yields

$$f(\boldsymbol{\gamma}) = P\mathbb{C}_{P-1}^{P-p_0}(e^{-\gamma_L})^{p_0}\left(1-e^{-\gamma_L}\right)^{P-p_0}\prod_{j=2}^{L}\delta(\gamma_j - \gamma_{j-1}). \tag{5.28}$$

Accordingly, the first expectation item in Eq. (5.26) can be approximated as

$$\begin{aligned}
&\frac{1}{12}\int_0^{\infty}\cdots\int_0^{\infty} f(\boldsymbol{\gamma})\left\{\exp\left(-\left(\left\|\boldsymbol{R}(\boldsymbol{x}_k - \hat{\boldsymbol{x}}_k)\right\|^2\right)^2\Big/2\sigma_k^2\right)\right\}d\gamma_1\cdots d\gamma_L\\
&=\frac{P\mathbb{C}_{P-1}^{P-p_0}}{12}\int_0^{\infty} e^{-(\frac{N_bN_m}{4\sigma^2 P}\sum_{l=1}^{L}\mathbb{E}[\beta_l]|\Delta x_{k,l}|^2+p_0)\gamma_L}(1-e^{-\gamma_L})^{P-p_0}d\gamma_L\\
&\overset{(c)}{=}\frac{P\mathbb{C}_{P-1}^{P-p_0}}{12}\mathbb{B}\left(\frac{N_bN_m}{4\sigma^2 P}\sum_{l=1}^{L}\mathbb{E}[\beta_l]\left|\Delta x_{k,l}\right|^2 + p_0, P - p_0 + 1\right)\\
&\overset{(d)}{=}\frac{P\mathbb{C}_{P-1}^{\nu-1}}{12}\mathbb{B}\left(\frac{N_bN_m}{4\sigma^2 P}\sum_{l=1}^{L}\mathbb{E}[\beta_l]\left|\Delta x_{k,l}\right|^2 + P + 1 - \nu, \nu\right),
\end{aligned} \tag{5.29}$$

where (c) comes from $\mathbb{B}(X, Y) = \int_0^1 \gamma^{X-1}(1-\gamma)^{Y-1}d\gamma$ [27]. (d) comes from $\nu = P - p_0 + 1$. Leveraging $p_{rc}(x) \approx 1/2\cos(\pi x/T_s) + 1/2$, $x \in (-T_s, T_s)$, we get

$$\mathbb{E}[\beta_l] = \int_0^{(N_c-1)T_s} \beta_l \frac{1}{(N_c-1)T_s} d\tau_l, \forall l. \tag{5.30}$$

At high SNR, we have

$$\begin{aligned}
&\frac{P\mathbb{C}_{P-1}^{\nu-1}}{12}\frac{(\frac{N_bN_m}{4\sigma^2 P}\sum_{l=1}^{L}\mathbb{E}[\beta_l]\left|\Delta x_{k,l}\right|^2 + P - \nu)!(\nu-1)!}{(\frac{N_bN_m}{4\sigma^2 P}\sum_{l=1}^{L}\mathbb{E}[\beta_l]\left|\Delta x_{k,l}\right|^2 + P)!}\\
&\simeq M_0\left(\frac{S}{N_0}\right)^{-\nu} + o\left\{\left(\frac{S}{N_0}\right)^{-\nu}\right\},
\end{aligned} \tag{5.31}$$

where M_0 is a constant irrelevant to S/N_0. Therefore ν ($1 \leq \nu \leq P$) represents the diversity gain [18].

(2) *Energy Efficiency*
Recall that in wGBM systems, the reduction in RF chains comes at the cost of more RF switches used for digital-to-analog module. To clarify such a cost is worthwhile, another important indicator, i.e., EE, is also included for comparison. Specifically, EE is defined as the ratio of SE and the MS-end total power consumption including the hardware part P_H and the communication part P_T. According to [28, 29], EE of wGBM can be calculated as

$$\begin{aligned}\varepsilon &= \frac{\eta}{P_T + P_H} \\ &= \frac{K\left(\left\lfloor \log_2 \mathbb{C}_L^{M_m} \right\rfloor + R\log_2 M\right)}{(K+L_{cp})\,(P_T+M_m P_{RF}+3M_m P_{SW})},\end{aligned} \tag{5.32}$$

where $P_{RF} = P_M + P_{LO} + P_{LPF} + P_{BBamp} + 2P_{DAC}$ represents the power consumption consumed by a single RF chain. The detailed definition and the power consumption in relation to RF chains are presented in Table 5.2. Particularly, P_{DAC} increases exponentially with the byte-width m, and linearly with the system bandwidth B and the DAC Walden's figure of merit c (the energy consumption per conversion step per Hz) (Table 5.2).

5.2.5 wGBM Accommodating Doppler

The above wGBM transceiver is designed in the absence of Doppler. To enhance its robustness against Doppler, in this section, we will introduce a first-order Doppler compensator at the BS without affecting the generic design.

In time-varying channels, the ICI may severely compromise the error performance. To address this issue, an additional operation, i.e., the first-order Doppler compensation, will be applied to the received signal $\mathbf{y}[n]$ (see Fig. 5.6) before removing CP. Such an receiver operation does not change the existing wGBM design and will demand negligible computational complexity.

Table 5.2 The definition and value corresponding to the notation defined in Eq. (5.32)

Notation	Definition	Value
P_{SW}	RF switches	5 mW
P_M	Mixer	16.8 mW
P_{LO}	Local oscillator	5 mW
P_{LPF}	Low pass filter	14 mW
P_{BBamp}	Based-band amplifier	5 mW
P_{DAC}	Digital-to-analog converter (DAC)	$cB2^m$

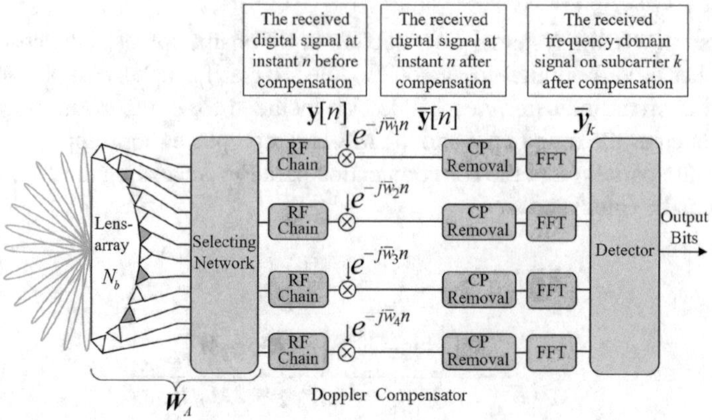

Fig. 5.6 The schematic of the Doppler compensator at the receiver end

Define $\bar{\mathbf{H}}_d[n] = \boldsymbol{W}_A^* \mathbf{H}_d[n] \boldsymbol{P}_A$, then the compensated signal for OFDM demodulator becomes

$$\bar{\mathbf{y}}[n] = \sum_{d=0}^{N_c-1} \mathbf{diag}\left\{e^{-j\bar{w}_l n}\right\}_{l=1}^{L} \bar{\mathbf{H}}_d[n]\mathbf{x}[n-d] + \bar{\mathbf{v}}[n], \tag{5.33}$$

where $\bar{w}_l$ stands for the Doppler associated with the l-th selected beam, and $\bar{\mathbf{v}}[n] \sim \mathcal{CN}\left(\mathbf{0}, \sigma^2 \boldsymbol{I}_L\right)$. The following two facts guarantee the effectiveness of the proposed first-order Doppler compensator.

Lemma 1 *For two beams whose AoAs (BS side) are θ_1 and θ_2, respectively, once satisfying $|\sin(\theta_2) - \sin(\theta_1)| > 2/N_b$, these two beams are approximately orthogonal.*

Proof Let $\Delta = \sin(\theta_2) - \sin(\theta_1)$, then the correlation between two beams can be calculated as

$$\left|\boldsymbol{a}_r(\theta_1)^* \boldsymbol{a}_r(\theta_2)\right| = \frac{1}{N_b}\left|\frac{\sin(\pi N_b \Delta/2)}{\sin(\pi \Delta/2)}\right|. \tag{5.34}$$

In mMIMO setup, the correlation decreases rapidly as $|\Delta|$ increases. Typically, a mmWave BS would equip many antennas (up to several hundred), so two beams coming from different directions tend to be orthogonal, resulting in a tiny interference regardless of Doppler compensation.

Lemma 2 *For two beams whose AoAs (BS side) are θ_1 and θ_2, respectively, once satisfying $|\sin(\theta_2) - \sin(\theta_1)| < 2/N_b$, the two beams experience a similar Doppler effect.*

Proof The relative Doppler shift Δw for these two beams can be calculated as

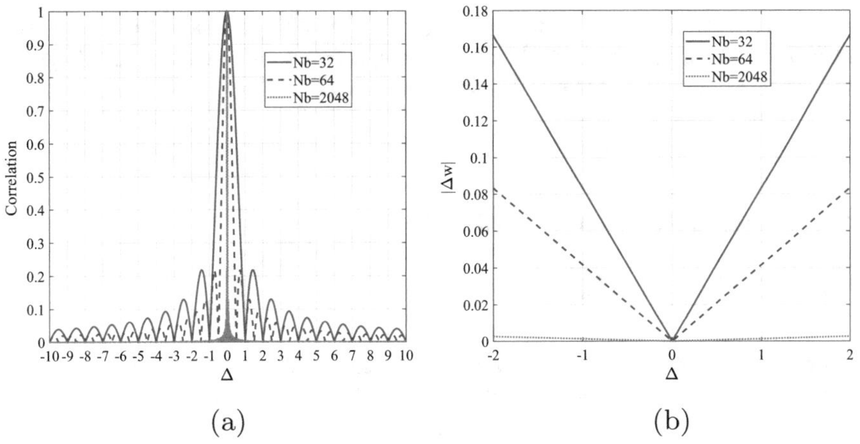

Fig. 5.7 The beam correlation and the Doppler difference versus the beam index gap. (**a**) Correlation of $\boldsymbol{a}_r(\theta_1)$ and $\boldsymbol{a}_r(\theta_2)$ (the interval of Δ is 1/16). (**b**) $|\Delta w|$ of w_1 and w_2 ($v_m = 80$ km/h and the interval of Δ is $2/N_b$)

$$\left|\frac{w_2 - w_1}{w_2}\right| = \left|\frac{2\pi f_c v_m T_s(\sin(\theta_2) - \sin(\theta_1))/c_v}{2\pi f_c v_m T_s \sin(\theta_2)/c_v}\right|$$
$$= \left|\frac{\Delta}{\sin(\theta_2)}\right| < \left|\frac{2}{N_b \sin(\theta_2)}\right|. \tag{5.35}$$

In mMIMO setup, Eq. (5.35) can be very small as $w_2 \approx w_1$. To illustrate Lemmas 1 and 2, we plot the Eq. (5.34) and Eq. (5.35) in Fig. 5.7a and b, respectively.

With the above two lemmas, $\bar{\mathbf{y}}[n]$ in Eq. (5.33) can henceforth be approximated as

$$\bar{\mathbf{y}}[n] \simeq \sum_{d=0}^{N_c-1} \bar{\mathbf{H}}_d \mathbf{x}[n-d] + \bar{\mathbf{v}}[n]. \tag{5.36}$$

where $\bar{\mathbf{H}}_d = \boldsymbol{W}_A^* \mathbf{H}_d \boldsymbol{P}_A$. Unlike the original $\bar{\mathbf{H}}_d[n]$, $\bar{\mathbf{H}}_d$ here turns out to be irrespective of n. That means, if neglecting the minimal interference or Doppler residue, $\mathbf{x}[n]$ is essentially transmitting through a time-invariant channel $\bar{\mathbf{H}}_d$. As N_b goes to infinity, it can be readily verified that the above compensation leads to an exact static channel $\bar{\mathbf{H}}_d$, such that ICIs can be completely vanished.

Let $\bar{\mathbf{y}} = [\bar{\mathbf{y}}_0^T, \bar{\mathbf{y}}_1^T, \cdots, \bar{\mathbf{y}}_{K-1}^T]^T$ be a vector consisting of all the received frequency-domain signals after OFDM demodulator. Then the I-O relationship of wGBM with Doppler compensation over time-varing channels can be concisely approximated as

$$\bar{\mathbf{y}} \simeq \bar{\boldsymbol{H}} \boldsymbol{s} + \boldsymbol{\xi}. \tag{5.37}$$

where $\bar{\boldsymbol{H}}$ has been defined in Eq. (5.9). To perform detection in time-varyingchannels, one can simply use the block-based ML detector in Eq. (5.15) and the low-complexity MMSE detector in Eq. (5.17).

5.3 Extension to Multi-User Setup

5.3.1 Design Challenges

Multi-user mmWave communication will be essential to future mobile communications. Some representative applications include, but are not limited to, smart city, environmental monitoring and intelligent agriculture. As a result, developing wGBM for multi-user mmWave systems would be crucial.

There are some difficulties in realizing the multi-user extension. First, designing wideband multi-user (wMU) mmWave transceivers is no easy work because signal processing in the analog domain is constrained to be frequency-flat, and a common combiner has to be employed at the BS to process the mixed signals from multiple users [30]. In addition, the hybrid structure complicates the interference management due to a reduced signal space in digital domain. This work jointly addresses the BS hybrid combining design and BS multi-user detection. In [31], a beam-switching based spatial modulation has been designed, which can yield an enhanced error performance in mmWave MIMO uplink. The authors in [32] proposed a novel wideband hybrid combining design to improve performance gains, with its corresponding user grouping and scheduling scheme. However, both schemes fail to combine the IM-based methods and orthogonal frequency division multiplexing (OFDM) together. Recall that IM-based OFDM scheme is an promising approach in mmWave multi-user systems, hence we will work on ensuring wGBM to be applicable to multi-user scenarios.

5.3.2 System Description

Let us consider an uplink multi-user mobile scenario with U users communicating with a common BS as shown in Fig. 5.8. Each user (UE) equipped with M_m RF chains and N_m antennas sends $L(M_m \leq L \leq N_m)$ data streams per subcarrier to the BS equipped with $M_b(UL \leq M_b \leq N_b)$ RF chains and N_b antennas. For ease of representation, we set $M_b = UL$. Other parameters are consistent with single-user wGBM.

Let $s_{k,u}$ be the transmitted signal on subcarrier k for UE-$u(u = 0, \cdots, U-1)$. Same to single-user wGBM transmission processing, the processed signal then goes through the devised digital-to-analog module and finally be precoded by the analog precoder

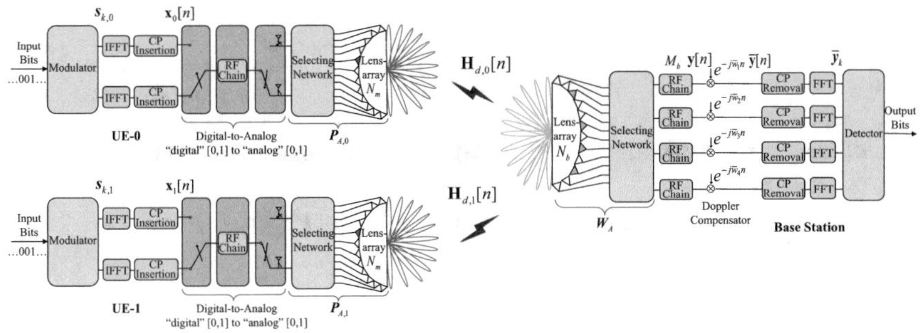

Fig. 5.8 Illustration of uplink wideband multi-user mmWave mMIMO system with $U = 2$, $M_m = 1$, $L = 2$, $M_b = 4$

$$\boldsymbol{P}_{A,u} = [\boldsymbol{p}_T(t_{1,u}), \boldsymbol{p}_T(t_{2,u}), \cdots, \boldsymbol{p}_T(t_{L,u})], \tag{5.38}$$

It is worth mentioning that $\boldsymbol{P}_{A,u}$ of different UEs are independent of each other.

We adopt a generic geometric channel model with N_c distinguishable delay taps and P dominant paths in order to exhibit the angular-delay sparsity and the doubly-selectivity of mmWave channels. The channel parameters from different users to the base station are distributed independently. For the channel in time domain for the UE-u, $u \in [0, U-1]$, the tap-d channel at time instant-n $\boldsymbol{H}_{d,u}[n] \in \mathcal{C}^{N_b \times N_m}$ can be written as:

$$\boldsymbol{H}_{d,u}[n] = \sqrt{\frac{N_b N_m}{P}} \sum_{p=1}^{P} \alpha_{p,u} p_{rc}(dT_s - \tau_{p,u}) \boldsymbol{a}_r(\theta_{p,u}) \boldsymbol{a}_t^*(\phi_{p,u}) e^{j w_{p,u} n} \tag{5.39}$$

For the p-th path of the UE-u, $\tau_{p,u}$ is the propagation delay uniformly distributed within $[0, (N_c-1)T_s)$; $w_{p,u} = 2\pi f_c v_{m,u} T_s \sin(\theta_{p,u})/c_v$ is its associated Doppler frequency shift; $\alpha_{p,u} \sim \mathcal{CN}(0,1)$ is the complex gain; $\theta_{p,u}$ and $\phi_{p,u}$ represent AoA and AoD, respectively. Then, under the setup of a half-wavelength spaced ULA, we have $\boldsymbol{a}_t(\phi_{p,u}) = \boldsymbol{f}_{N_m}(\pi \sin(\phi_{p,u}))$ and $\boldsymbol{a}_r(\theta_{p,u}) = \boldsymbol{f}_{N_b}(\pi \sin(\theta_{p,u}))$. Given the time-domain channel response, the corresponding inter-subcarrier channel response $\boldsymbol{H}_{k,u}[r] (k, r \in [0, K-1])$ can be expressed as

$$\boldsymbol{H}_{k,u}[r] = \frac{1}{K} \sum_{i=0}^{K-1} \sum_{d=0}^{N_c-1} \boldsymbol{H}_{d,u}[L_{cp} + i] e^{-j2\pi(rd + (k-r)i)/K}. \tag{5.40}$$

where $\boldsymbol{H}_{k,u}[r] \in \mathcal{C}^{N_b \times N_m}$ represents the channel response between the UE-u and BS at subcarrier-k if $k = r$, or the ICI from subcarrier-r to k if $k \neq r$.

Apparently, when $w_{p,u} = 0$, $\mathbf{H}_{d,u}[n]$ turns out to be time-invariant. Denote the static version of $\mathbf{H}_{d,u}[n]$ as $\mathbf{H}_{d,u},$. In this case, all the inter-subcarrier channels for UE-u vanish except for leaving

$$\boldsymbol{H}_{k,u}[k] = \boldsymbol{H}_{k,u} = \sum_{d=0}^{N_c-1} \mathbf{H}_{d,u} e^{-j\frac{2\pi k}{K}d}. \tag{5.41}$$

After channel propagation, the received signal on subcarrier k is given by

$$\boldsymbol{r}_k = \sum_{u=0}^{U-1} \boldsymbol{H}_{k,u}[k]\boldsymbol{P}_{A,u}\boldsymbol{s}_{k,u} + \sum_{u=0}^{U-1}\sum_{r=0,r\neq k}^{K-1} \boldsymbol{H}_{k,u}[r]\boldsymbol{P}_{A,u}\boldsymbol{s}_{r,u} + \boldsymbol{w}_k. \tag{5.42}$$

$\boldsymbol{r}_k$ is first combined by the analog combiner

$$\boldsymbol{W}_A = [\boldsymbol{f}_R(r_1), \boldsymbol{f}_R(r_2), \cdots, \boldsymbol{f}_R(r_{M_b})] \in \mathcal{C}^{N_b \times UL}, \tag{5.43}$$

with $\boldsymbol{f}_R(i)$ being the i-th column of $\boldsymbol{F}_{N_b}$. After CP removal and FFT operation, the signal to be processed in digital baseband becomes

$$\begin{aligned}\boldsymbol{y}_k &= \sum_{u=0}^{U-1} \mathbf{W}_A^* \boldsymbol{H}_{k,u}[k]\boldsymbol{P}_{A,u}\boldsymbol{s}_{k,u} + \\ &\sum_{u=0}^{U-1}\sum_{r=0,r\neq k}^{K-1} \mathbf{W}_A^* \boldsymbol{H}_{k,u}[r]\boldsymbol{P}_{A,u}\boldsymbol{s}_{r,u} + \mathbf{W}_A^*\boldsymbol{w}_k. \\ &= \sum_{u=0}^{U-1} \bar{\boldsymbol{H}}_{k,u}[k]\boldsymbol{s}_{k,u} + \sum_{u=0}^{U-1}\sum_{r=0,r\neq k}^{K-1} \bar{\boldsymbol{H}}_{k,u}[r]\boldsymbol{s}_{r,u} + \boldsymbol{\xi}_k,\end{aligned} \tag{5.44}$$

where $\boldsymbol{\xi}_k = \boldsymbol{W}_A^*\boldsymbol{w}_k \sim \mathcal{CN}\left(\mathbf{0}, \sigma^2 \boldsymbol{I}_{M_b}\right)$ is the combined noise.

Let $\boldsymbol{s}_u = \left[\mathbf{s}_{0,u}^T, \mathbf{s}_{2,u}^T, \cdots \mathbf{s}_{K-1,u}^T\right]^T$ $(u \in [0, U-1])$ being the transmitted signal from UE-u, then the system I-O relationship can be cast into

$$\underbrace{\begin{bmatrix} \boldsymbol{y}_0 \\ \vdots \\ \boldsymbol{y}_{K-1} \end{bmatrix}}_{\boldsymbol{y}} = \underbrace{[\boldsymbol{H}_0, \cdots, \boldsymbol{H}_{U-1}]}_{\boldsymbol{H}_U} \underbrace{\begin{bmatrix} \boldsymbol{s}_0 \\ \vdots \\ \boldsymbol{s}_{U-1} \end{bmatrix}}_{\boldsymbol{s}_U} + \underbrace{\begin{bmatrix} \boldsymbol{\xi}_0 \\ \vdots \\ \boldsymbol{\xi}_{K-1} \end{bmatrix}}_{\boldsymbol{\xi}} \tag{5.45}$$

with

$$\boldsymbol{H}_u = \begin{bmatrix} \bar{\boldsymbol{H}}_{0,u}[0] & \cdots & \bar{\boldsymbol{H}}_{0,u}[K-1] \\ \vdots & \ddots & \vdots \\ \bar{\boldsymbol{H}}_{K-1,u}[0] & \cdots & \bar{\boldsymbol{H}}_{K-1,u}[K-1] \end{bmatrix}. \tag{5.46}$$

Define $\bar{\boldsymbol{H}}_{k,u} = \boldsymbol{W}_A^* \boldsymbol{H}_{k,u} \boldsymbol{P}_{A,u}$, when $w_{p,u} = 0$, then $\boldsymbol{H}_U$ in Eq. (5.45) can be further expressed as

$$\bar{\boldsymbol{H}}_U = \left[\bar{\boldsymbol{H}}_0, \cdots, \bar{\boldsymbol{H}}_{U-1}\right], \tag{5.47}$$

with Eq. (5.46) being simplified as

$$\bar{\boldsymbol{H}}_u = \mathbf{diag}\left\{\bar{\boldsymbol{H}}_{k,u}\right\}_{k=0}^{K-1}. \tag{5.48}$$

5.3.3 wGBM wMU Transceiver in Static Channels

In this section, we detail the wideband multi-user transceiver design in the absence of Doppler. For each UE, the modulator design at the Tx is same to the wGBM Tx design, i.e., Eqs. (5.10)–(5.13). Specially, similar to Eq. (5.10), we can first obtain the corresponding data vector of UE-u on subcarrier-k as $\boldsymbol{x}_{k,u}$. After FFT module, a stacked frequency-domain signal vector of UE-u is obtained as

$$\boldsymbol{s}_U = \boldsymbol{I}_U \otimes (\boldsymbol{F}_K \otimes \boldsymbol{I}_L)\boldsymbol{x}_U, \tag{5.49}$$

with $\boldsymbol{x}_U$ being a concatenation of $\{\boldsymbol{x}_u\}_{u=0}^{U-1}$, and $\boldsymbol{x}_u$ being a concatenation of $\{\boldsymbol{x}_{k,u}\}_{k=0}^{K-1}$. Next, we will introduce the wideband multi-user receiver design and beam selection design in detail.

(1) ML detector: Based on Eqs. (5.11) and (5.45), the ML detector can be expressed as

$$\begin{aligned}
\hat{\boldsymbol{x}}_U &= \underset{\forall k, \boldsymbol{x}_{k,u} \in G}{\arg\min} \left\| \boldsymbol{y} - \bar{\boldsymbol{H}}_U \boldsymbol{s}_U \right\|^2 \\
&= \underset{\forall k, \boldsymbol{x}_{k,u} \in G}{\arg\min} \left\| \boldsymbol{y} - \bar{\boldsymbol{H}}_U (\boldsymbol{I}_U \otimes (\boldsymbol{F}_K \otimes \boldsymbol{I}_L))\boldsymbol{x}_U \right\|^2,
\end{aligned} \tag{5.50}$$

where G contains all possible $\boldsymbol{x}_{k,u}$'s. Since the detection is performed across all subcarrier symbols and users, the computational complexity grows exponentially with KU. It also drives us to seek a low-complexity alternative for practice use.

(2) LMMSE-based detector: Similar to single user, to reduce the complexity, one has to decouple the intertwined relationship among all subcarriers and users. Towards this objective, a two-step low-complexity detector is proposed. First, we apply the widely-used LMMSE equalizer $\boldsymbol{M}_U = \{\bar{\boldsymbol{H}}_U^* \bar{\boldsymbol{H}}_U + \sigma^2 \boldsymbol{I}_{LK}\}^{-1} \bar{\boldsymbol{H}}_U^*$ to get an estimated $\boldsymbol{s}_U$, i.e., $\bar{\boldsymbol{s}}_U = \boldsymbol{M}_U \boldsymbol{y}$. Leveraging Eq. (5.49), $\boldsymbol{x}_k$ can accordingly be estimated as

$$\begin{aligned}\bar{\boldsymbol{x}}_{k,u} &= \boldsymbol{I}_U \otimes (\boldsymbol{F}_K^*[k,:] \otimes \boldsymbol{I}_L)\bar{\boldsymbol{s}}_U \\ &= \boldsymbol{I}_U \otimes (\boldsymbol{F}_K^*[k,:] \otimes \boldsymbol{I}_L)\boldsymbol{M}_U \boldsymbol{y}. \end{aligned} \tag{5.51}$$

Next, the ML detection is applied to the low-dimensional vector $\bar{\boldsymbol{x}}_{k,u}$, giving rise to the demodulated data vector

$$\hat{\boldsymbol{x}}_{k,u} = \arg\min_{\boldsymbol{x}_{k,u} \in G} \left\| \bar{\boldsymbol{x}}_{k,u} - \boldsymbol{x}_{k,u} \right\|^2. \tag{5.52}$$

(3) AMP-based detector: Similar to single user wGBM system, the AMP-based detector aims to decouple a complicated KLU-dimension joint estimation problem into KLU simple ones, each dealing with a scalar estimation. Specially, we just need to replace $\tilde{\boldsymbol{H}}$ with $\tilde{\boldsymbol{H}}_U = \bar{\boldsymbol{H}}_U(\boldsymbol{F}_K \otimes \boldsymbol{I}_L)$ in Algorithm 3. Similarly, define $P_u = \bigcup_{\forall k} P_{k,u}$ and set $(\boldsymbol{r}_u, \boldsymbol{t}_u) = \left\{(r_{1,u}, t_{1,u}), \cdots, (r_{L,u}, t_{L,u})\right\} \subset \mathcal{P}_u$ with $\forall i$, $(r_{i,u}, t_{i,u}) \in \mathcal{P}_U$. Based on the selection criterion of minimizing the mean square error, $\|\boldsymbol{e}_U\|^2 = \|\bar{\boldsymbol{s}}_U - \boldsymbol{s}_U\|^2 = \|\boldsymbol{M}_U \boldsymbol{y} - \boldsymbol{s}_U\|^2$, the LU beam indices for constructing the analog beamformer can be determined as

$$\begin{aligned}&\left\{(\bar{r}_{1,0}, \bar{t}_{1,0}), \cdots, (\bar{r}_{L,u}, \bar{t}_{L,u}), \cdots, (\bar{r}_{L,U-1}, \bar{t}_{L,U-1})\right\} \\ &= \arg\min_{\forall(\boldsymbol{r}_u, \boldsymbol{t}_u) \subset \mathcal{P}_u} Tr\left\{\{\bar{\boldsymbol{H}}_U^* \bar{\boldsymbol{H}}_U\}^{-1}\right\}. \end{aligned} \tag{5.53}$$

The analog precoder of UE-u and combiner of the BS are accordingly designed as

$$\boldsymbol{P}_{A,u} = [\boldsymbol{f}_T(\bar{t}_{1,u}), \boldsymbol{f}_T(\bar{t}_{2,u}), \cdots, \boldsymbol{f}_T(\bar{t}_{L,u})], \tag{5.54a}$$

$$\boldsymbol{W}_A = [\boldsymbol{f}_R(\bar{r}_{1,0}), \cdots, \boldsymbol{f}_R(\bar{r}_{l,u}), \cdots, \boldsymbol{f}_R(\bar{r}_{L,U-1})]. \tag{5.54b}$$

5.3.4 wGBM wMU Accommodating Doppler

The above wMU uplink transceiver is designed in the absence of Doppler. In time-varying channels, the first-order Doppler compensation will be applied to the received signal $\mathbf{y}[n]$ (see Fig. 5.8) before removing CP. Define $\bar{\mathbf{H}}_{d,u}[n] = \boldsymbol{W}_A^* \mathbf{H}_{d,u}[n] \boldsymbol{P}_{A,u}$. Then the compensated signal for OFDM demodulator becomes

$$\bar{\mathbf{y}}[n] = \sum_{u=0}^{U-1} \sum_{d=0}^{N_c-1} \mathbf{diag}\left\{e^{-j\bar{\omega}_l n}\right\}_{l=1}^{M_b} \bar{\mathbf{H}}_{d,u}[n]\mathbf{x}_u[n-d] + \bar{\mathbf{v}}[n], \tag{5.55}$$

where $\bar{\omega}_l$ stands for the Doppler associated with the l-th selected beam, and $\bar{\mathbf{v}}[n] \sim \mathcal{CN}\left(\mathbf{0}, \sigma^2 \boldsymbol{I}_{M_b}\right)$. With the above two lemmas in Eqs. (5.34) and (5.35), $\bar{\mathbf{y}}_U[n]$ in Eq. (5.55) can henceforth be approximated as

$$\bar{\mathbf{y}}[n] \simeq \sum_{u=0}^{U-1} \sum_{d=0}^{N_c-1} \bar{\mathbf{H}}_{d,u} \mathbf{x}_u[n-d] + \bar{\mathbf{v}}[n]. \tag{5.56}$$

where $\bar{\mathbf{H}}_{d,u} = \boldsymbol{W}_A^* \mathbf{H}_{d,u} \boldsymbol{P}_{A,u}$.

Let $\bar{\boldsymbol{y}}_U = [\bar{\boldsymbol{y}}_0^T, \bar{\boldsymbol{y}}_1^T, \cdots, \bar{\boldsymbol{y}}_{K-1}^T]^T$ be a vector consisting of all the received frequency-domain signals after OFDM demodulator. Then the I-O relationship of wideband multi-user uplink system with Doppler compensation over time-varying channels can be concisely approximated as

$$\bar{\boldsymbol{y}} \simeq \bar{\boldsymbol{H}}_U \boldsymbol{s}_U + \boldsymbol{\xi}. \tag{5.57}$$

where $\bar{\boldsymbol{H}}_U$ has been defined in Eq. (5.47). To perform detection in time-varying channels, one can simply use the block-based ML detector in Eq. (5.50) and the low-complexity MMSE detector in Eq. (5.52).

5.4 Simulations

In this section, we first compare our proposed wGBM with the conventional spatial multiplexing (CSM) and the maximum beamforming (MBF) for $U = 1$ in terms of EE and BER performance. CSM uses the selected beams without subcarrier-based IM for transmission, while MBF uses the strongest beam for transmission. Then, we extend the corresponding wGBM multi-user extension scenario for $U = 3$. The related system and channel parameters for simulation are provided in Table 5.3. The average SNR is defined as $\frac{S}{N_0} = \frac{P_T/\alpha}{\sigma^2} = \frac{P_T/\alpha}{n_0 B}$, where α is the path loss for a certain transmission distance R_s. We further investigate the proposed wMU scheme using a realistic slow-mobility non-stationary 5G beyond vehicular mmWave mMIMO channels.

5.4.1 *Energy Efficiency*

In Fig. 5.9, we compare the EE performance among wGBM, CSM and MBF by setting $U = 1$. It is worth noting that MBF only employs one RF chains at the MS, which does not involve the variation of RF chains. We set $P_T = 1$W. The MS-end hardware power consumption P_H for CSM and MBF are $M_m(P_{RF} + P_{SW})$ and P_{RF}, respectively. Besides, we consider three modulation orders: $M = 4$, $M = 8$ and $M = 32$, with $L = 8$. From Fig. 5.9, we have the following observations:

Table 5.3 The parameters for simulation

N_m	32
N_b	32 or 64
R_s	100m
f_c	60GHz
K	256
N_c	64
L_{cp}	64
P	10
B	1GHz
m	9bits
c	494fJ/step/Hz
ψ	$U[-\frac{7\pi}{16}, \frac{7\pi}{16}]$
$p_{rc}(\cdot)$	Raise cosine with $\beta = 0.8$
Noise power spectral density n_0	-174 dBm/Hz

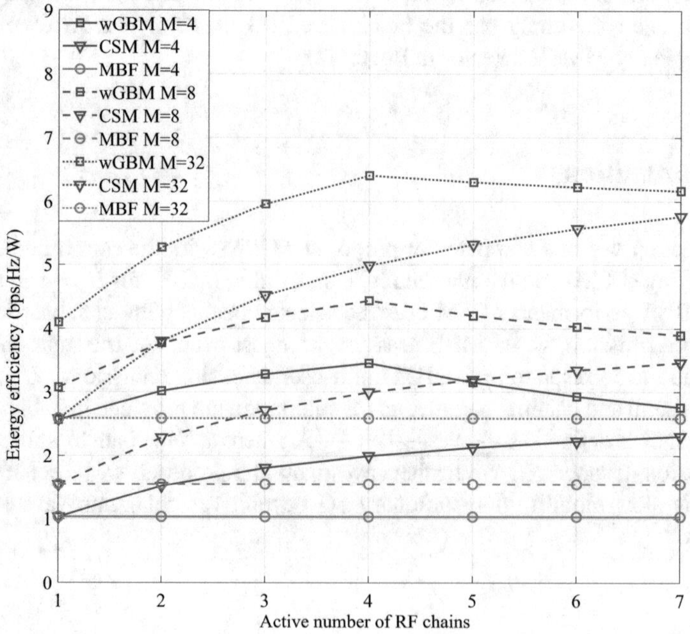

Fig. 5.9 EE comparisons among wGBM, CSM, and MBF

- With the same number of RF chains, wGBM is always superior to CSM because the former can exploit a higher MG.
- Despite of smallest power consumption by employing one RF chain, MBF is still inferior to wGBM and CSM because the former has the lowest MG.

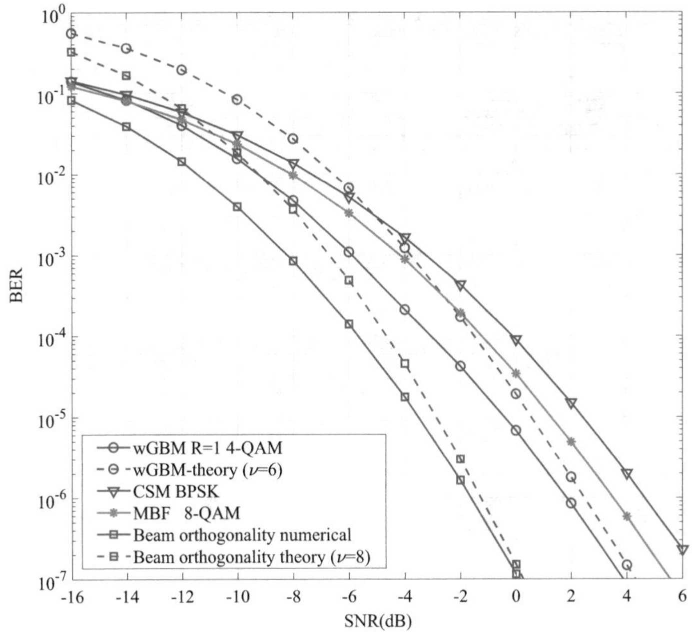

Fig. 5.10 BER comparisons among wGBM, CSM, and MBF in FSTI channels ($\eta = 2.4$ bps/Hz, $N_m = 32$ and $N_b = 32$)

- When the number of the active RF chains exceeds a certain value, wGBM may experience an EE decline because the benefit of a higher SE arising from more RF chains cannot compensate the additional power consumption.

5.4.2 Error Performance in Doubly-Selective Channels

In Fig. 5.10, we make BER comparisons among wGBM, CSM and MBF in frequency-selective time-invariant (FSTI) channels with a SE of 2.4 bps/Hz. wGBM related parameters are set as $L = 3$, $M_m = 1$ with 4-QAM modulation. CSM and MBF adopt BPSK and 8-QAM modulation, respectively. With SNR increases, the advantage of wGBM over CSM and MBF gradually becomes noticeable. At high SNR, wGBM achieves a coding gain of 2.5 and 1.5 dB over CSM and MBF, respectively. These advantages are owing to a 33% ratio of index bits, which are more robust in high SNR region.

In Fig. 5.11, we consider a higher SE by setting $L = 3$, $M_m = 2$, and 8-QAM for wGBM. To achieve the same SE of 5.6 bps/Hz, CSM adopts hybrid 4-QAM and 8-QAM modulation, while MBF adopts 128-QAM modulation. At low SNR, we observe that wGBM is inferior to CSM because the index bits are susceptible

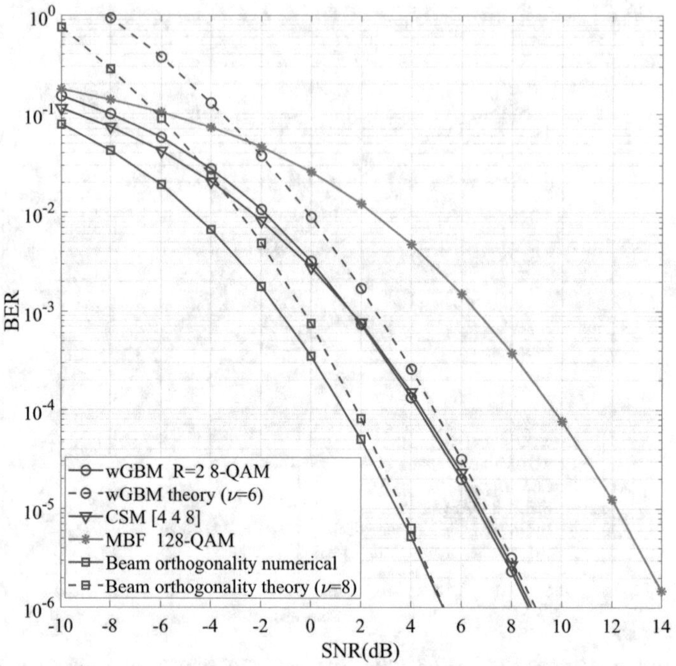

Fig. 5.11 BER comparisons among wGBM, CSM, and MBF in FSTI channels ($\eta = 5.6$ bps/Hz, $N_m = 32$ and $N_b = 32$)

to strong noises and thereby leading to more frequent erroneous detection. At high SNR, wGBM performs similar to CSM because the detection loss of wGBM caused by the MMSE detector can partially be bridged by the advantage of index modulation. With 14% index bits, wGBM also enjoys a coding gain of 6 dB over MBF. Since MBF adopts a high-order (128) modulation, it performs worst. For wGBM, we further compare the theoretical and numerical results under perfect and imperfect beam orthogonality, and we find both are well matched at high SNR.

Following the above system configuration, we then compare the LMMSE-based and AMP-based detectors with SE equal to 2.4 bps/Hz and 5.6 bps/Hz in Fig. 5.12. When applying AMP-based detection, we set $T_{max} = 5$ and $\delta = 0.4$. Simulations show that the AMP-based detector achieves a coding gain of about 0.5 dB over the LMMSE-based detector at BER = 10^{-6}. This is because the latter algorithm can well explore the structural sparsity of the transmitted signal. Although yielding a lower BER, this one is unfriendly in guiding beam selection as the LMMSE-based detector does for the difficulty in obtaining a closed-form expression.

We then test whether the BER advantage still holds in frequency-selective time-varying(FSTV) channels in Figs. 5.13 and 5.14. The performance of wGBM in FSTI channels is set as the ideal benchmark. The modulation parameters remain identical to those adopted in Fig. 5.10. For frequency-selective time-varying(FSTV)

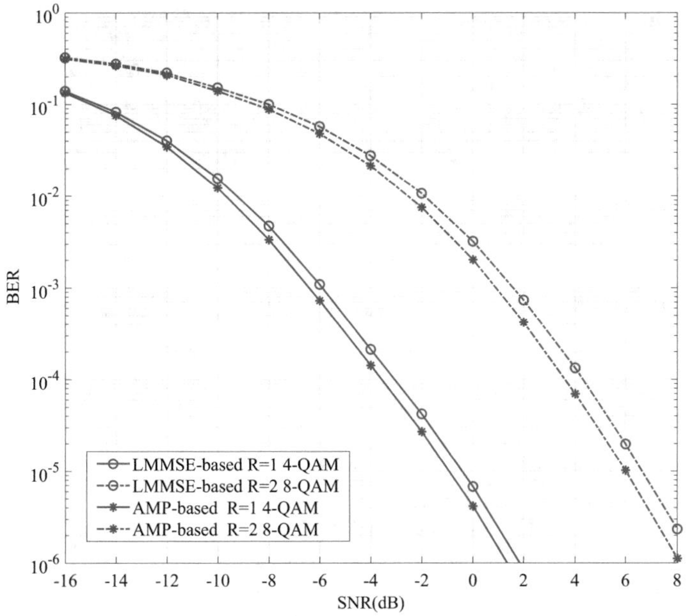

Fig. 5.12 BER comparisons of wGBM with different detector ($\eta = 2.4$ bps/Hz, 5.6 bps/Hz, $N_m = 32$ and $N_b = 32$)

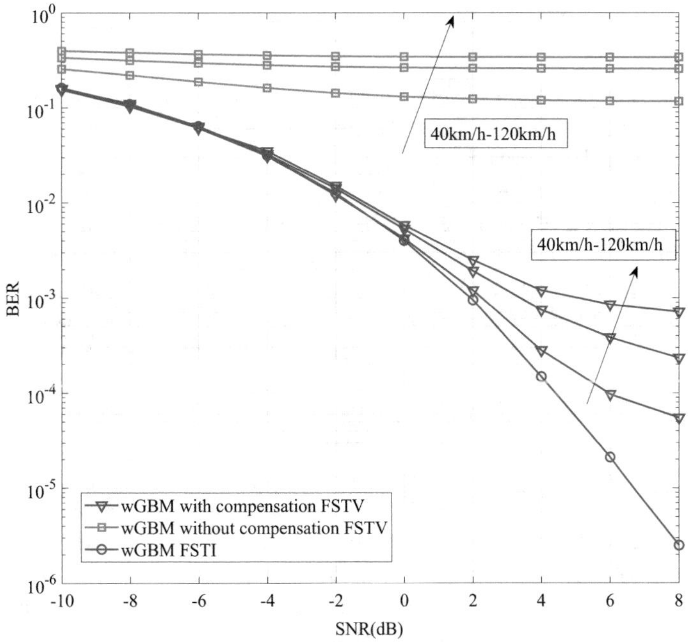

Fig. 5.13 BER comparisons in doubly-selective channels ($\eta = 5.6$ bps/Hz, $N_m = 32$ and $N_b = 32$)

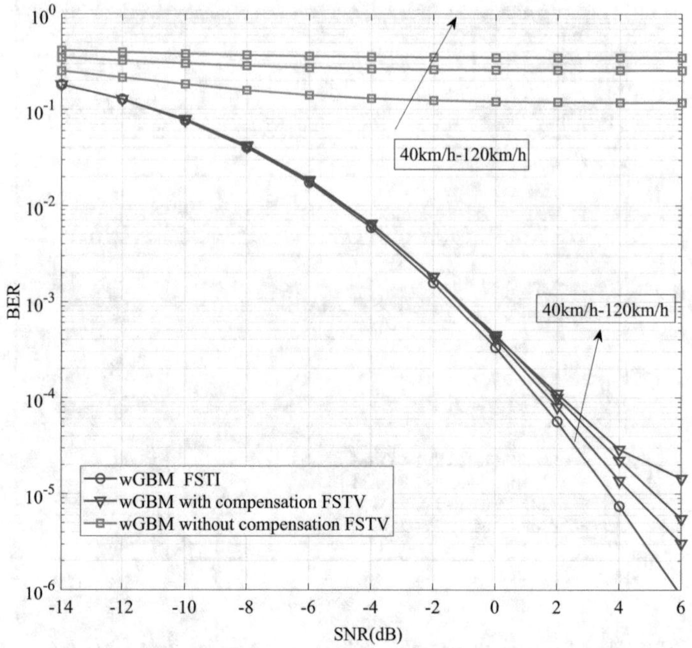

Fig. 5.14 BER comparisons in doubly-selective channels (η = 5.6 bps/Hz, N_m = 32 and N_b = 64)

channels, we consider three different levels of maximum Doppler shift: 0.005, 0.01, and 0.015, corresponding to v_m = 40 km/h, 80 km/h, and 120 km/h, respectively. As can be seen from Figs. 5.13 and 5.14, wGBM with compensation performs dramatically better than that without compensation. In Fig. 5.13, the gap between the compensated wGBM and the ideal benchmark is minimal at relatively low SNR; while in the high SNR region, the former exhibits a BER floor as v_m increases. However, this error floor can be lowered by increasing N_b. As shown in Fig. 5.14, when v_m = 120 km/h, the error floor is reduced from 10^{-3} to 10^{-5} as N_b increases from 32 to 64. This is because the latter configuration enjoys a higher angular resolution, therefore leaving a smaller interference residue after compensation.

We then consider a multi-user setup by setting $U = 3$. For each user, wGBM is set as $L = 2$, $M_m = 1$ with 4-QAM modulation. CSM and MBF adopt BPSK and 8-QAM modulation, respectively. First, we compare wGBM wMU scheme with its corresponding counterparts with LMMSE detector in Fig. 5.15. Similar to sing-user wGBM, at low SNR, we observe that wGBM wMU is inferior to CSM, while at high SNR, wMU wGBM outperforms CSM. To test whether the wMU BER advantage still holds in FSTV channels, more simulations are made with the results shown in Fig. 5.16. The performance of wGBM wMU in FSTI channels is set as the ideal benchmark. Here, we also consider three different levels of maximum Doppler shift:

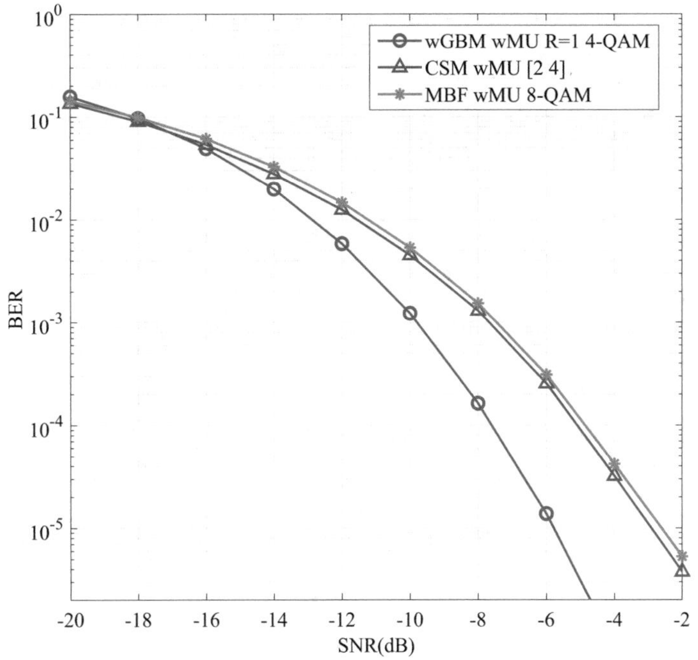

Fig. 5.15 BER comparisons among wGBM, CSM, and MBF in FSTI channels ($U = 3$, $\eta = 2.4$ bps/Hz, $L = 2$, $N_m = 32$ and $N_b = 64$)

0.005, 0.01, and 0.015, corresponding to $v_m = 40$ km/h, 80 km/h, and 120 km/h, respectively. We observe that wGBM wMU also performs well at three levels of mobility.

5.4.3 More Test in Low Vehicular Traffic Density (VTD) Slow-Mobility Non-stationary Channels

We further test the proposed wGBM wMU scheme using a low VTD slow-mobility non-stationary beyond 5G and 6G vehicular mmWave mMIMO channels [33] in Fig. 5.17. We observe that wGBM also outperforms others. The maintenance of such an advantage is because the proposed scheme does not impose much constraint on the channel itself. Therefore, varying the AoAs/AoDs distribution or beam direction does not affect its feasibility.

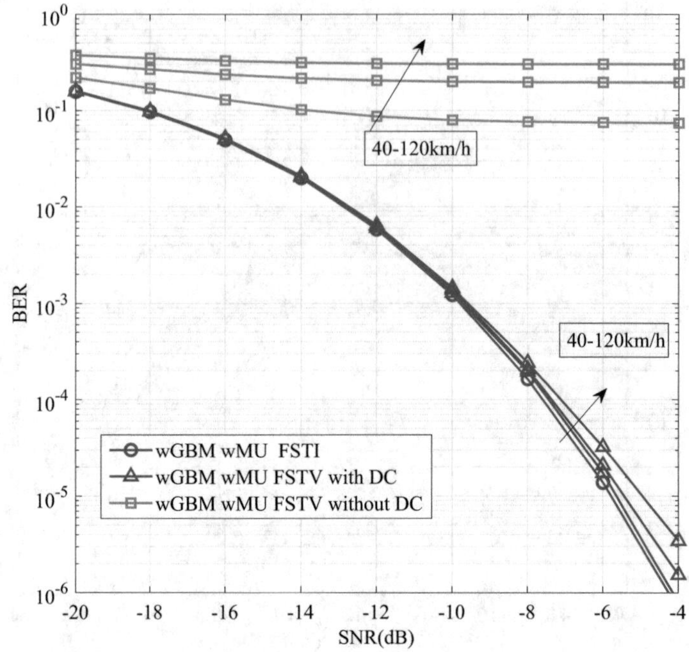

Fig. 5.16 BER comparisons of wGBM in doubly-selective channels ($U = 3$, $\eta = 2.4$ bps/Hz, $L = 2$, $N_m = 32$ and $N_b = 64$)

5.5 Discussions and Summary

In this section, we studied index modulation for mmWave vehicular uplink access. First, we generalize the SE-enhanced GBM scheme from the narrowband to realistic wideband scenarios. The resultant design is termed as wGBM. With the help of a novel symbol-based IM framework, the developed wGBM is compatible with hybrid OFDM systems while maintaining the SE-enhanced merit of GBM. To avoid the high detection complexity in large-scale OFDM systems, we devise a low-complexity detector by transforming the complicated symbol (block) detection into the manageable subcarrier (sub-block) detection. Apart from the generic wGBM transceiver design, a simple first-order Doppler compensator is carefully designed to enhance the robustness of wGBM against Doppler. To support massive connection and ultra-fast speed for the next-generation cellular, we extend the single user wGBM to wMU applications. Theoretical analyses and numerical simulations have been carried out to validate the advantages of wGBM in terms of BER and EE.

For mmWave vehicular uplink access, the number of RF chains limits the number of individual streams that a common BS can simultaneously serve. Thus, for future study, it is essential to design an optimization scheduling mechanism that jointly considers user selection, beam pairing, and frequency resource allocation for the different optimization goals of uplink vehicular wMU mmWave MIMO systems.

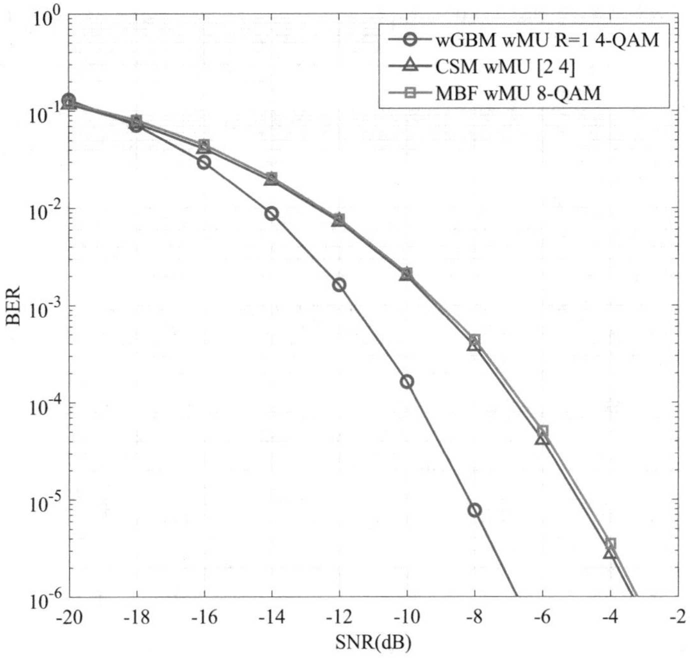

Fig. 5.17 BER comparisons among wGBM, CSM, and MBF in low VTD slow-mobility Non-stationary beyond 5G and 6G vehicular mmWave mMIMO channels ($U = 3$, $\eta = 2.4$ bps/Hz, $L = 2$, $N_m = 32$ and $N_b = 64$)

References

1. X. Cheng, M. Zhang, M. Wen, L. Yang, Index modulation for 5G: striving to do more with less. IEEE Wirel. Commun. **25**(2), 126–132 (2018)
2. R. Mesleh, H. Haas, S. Sinanovic, C.W. Ahn, S. Yun, Spatial modulation. IEEE Trans. Veh. Technol. **57**(4), 2228–2241 (2008)
3. M. Di Renzo, H. Haas, A. Ghrayeb, S. Sugiura, L. Hanzo, Spatial modulation for generalized MIMO: challenges, opportunities, and implementation. Proc. IEEE **102**(1), 56–103 (2014)
4. E. Basar, U. Aygolu, E. Panayirci, H.V. Poor, Orthogonal frequency division multiplexing with index modulation. IEEE Trans. Signal Proc. **61**(22), 5536–5549 (2013)
5. A. Younis, N. Serafimovski, R. Mesleh, H. Haas, Generalised spatial modulation, in *2010 Conference Record of the Forty Fourth Asilomar Conference on Signals, Systems and Computers* (2010), pp. 1498–1502
6. Y. Bian, X. Cheng, M. Wen, L. Yang, H.V. Poor, B. Jiao, Differential spatial modulation. IEEE Trans. Veh. Technol. **64**(7), 3262–3268 (2015)
7. S. Gao, M. Zhang, X. Cheng, Precoded index modulation for multi-input multi-output OFDM. IEEE Trans. Wirel. Commun. **17**(1), 17–28 (2018)
8. Y. Fan, Y. Jia, W. Qu, X. Cheng, X. Mu, L. Yang, Differential spatial frequency modulation with orthogonal frequency division multiplexing, in *2018 IEEE Global Communications Conference (GLOBECOM)* (2018), pp. 1–6
9. F. Rusek, D. Persson, B.K. Lau, E.G. Larsson, T.L. Marzetta, O. Edfors, F. Tufvesson, Scaling up MIMO: Opportunities and challenges with very large arrays. IEEE Signal Process. Mag. **30**(1), 40–60 (2013)

10. F. Boccardi, R.W. Heath, A. Lozano, T.L. Marzetta, P. Popovski, Five disruptive technology directions for 5G. IEEE Commun. Mag. **52**(2), 74–80 (2014)
11. A.L. Swindlehurst, E. Ayanoglu, P. Heydari, F. Capolino, Millimeter-wave massive MIMO: the next wireless revolution? IEEE Commun. Mag. **52**(9), 56–62 (2014)
12. J. Brady, N. Behdad, A.M. Sayeed, Beamspace MIMO for millimeter-wave communications: system architecture, modeling, analysis, and measurements. IEEE Trans. Antennas and Propag. **61**(7), 3814–3827 (2013)
13. S. Han, C. I, Z. Xu, and C. Rowell, Large-scale antenna systems with hybrid analog and digital beamforming for millimeter wave 5G. IEEE Commun. Mag. **53**(1), 186–194 (2015)
14. F. Sohrabi, W. Yu, Hybrid digital and analog beamforming design for large-scale antenna arrays. IEEE J. Selec. Topics Signal Process. **10**(3), 501–513 (2016)
15. L. He, J. Wang, J. Song, Spatial modulation for more spatial multiplexing: Rf-chain-limited generalized spatial modulation aided mm-wave MIMO with hybrid precoding. IEEE Trans. Commun. **66**(3), 986–998 (2018)
16. Y. Ding, K.J. Kim, T. Koike-Akino, M. Pajovic, P. Wang, P. Orlik, Spatial scattering modulation for uplink millimeter-wave systems. IEEE Commun. Lett. **21**(7), 1493–1496 (2017)
17. Y. Ding, V. Fusco, A. Shitvov, Y. Xiao, H. Li, Beam index modulation wireless communication with analog beamforming. IEEE Trans. Vehic. Technol. **67**(7), 6340–6354 (2018)
18. S. Gao, X. Cheng, L. Yang, Spatial multiplexing with limited RF chains: Generalized beamspace modulation (GBM) for mmwave massive MIMO. IEEE J. Selec. Areas Commun. **37**(9), 2029–2039 (2019)
19. J. Zhang, Y. Huang, J. Wang, L. Yang, Hybrid precoding for wideband millimeter-wave systems with finite resolution phase shifters. IEEE Trans. Vehic. Technol. **67**(11), 11285–11290 (2018)
20. F. Sohrabi, W. Yu, Hybrid analog and digital beamforming for mmwave OFDM large-scale antenna arrays. IEEE J. Selec. Areas Commun **357**, 1432–1443 (2017)
21. W. Shen, X. Bu, X. Gao, C. Xing, L. Hanzo, Beamspace precoding and beam selection for wideband millimeter-wave MIMO relying on lens antenna arrays. IEEE Trans. Signal Process. **67**(24), 6301–6313 (2019)
22. Y. Fan, S. Gao, X. Cheng, L. Yang, N. Wang, Wideband generalized beamspace modulation (wGBM) for mmwave massive MIMO over doubly-selective channels. IEEE Trans. Vehic. Technol. **70**(7), 6869–6880 (2021)
23. J. Li, S. Gao, X. Cheng, L. Yang, Hybrid precoded spatial modulation (hPSM) for mmwave massive MIMO systems over frequency-selective channels. IEEE Wirel. Commun. Lett. **9**(6), 839–842 (2020)
24. S. Gao, X. Cheng, L. Yang, Estimating doubly-selective channels for hybrid mmwave massive MIMO systems: a doubly-sparse approach. IEEE Trans. Wirel. Commun. **19**(9), 5703–5715 (2020)
25. L. Rugini, P. Banelli, G. Leus, OFDM Communications over time-varying channels, in *Wireless Communications Over Rapidly Time-Varying Channels*, ed. by F. Hlawatsch, G. Matz (Academic, New York, 2011)
26. L. Wei, J. Zheng, Q. Liu, Approximate message passing detector for index modulation with multiple active resources. IEEE Trans. Vehic. Technol. **68**(1), 972–976 (2019)
27. I.S. Gradshteyn, I.M. Ryzhik, *Table of Integrals, Series, and Products* (Elsevier, San Diego, 2007)
28. X. Gao, L. Dai, A.M. Sayeed, Low RF-complexity technologies to enable millimeter-wave MIMO with large antenna array for 5G wireless communications. IEEE Commun. Mag. **56**(4), 211–217 (2018)
29. S. Gao, X. Cheng, L. Fang, L. Yang, Model enhanced learning based detectors (Me-LeaD) for wideband multi-user 1-bit mmWave communications. IEEE Trans. Wirel. Commun. **20**(7), 4646–4656 (2021)
30. X. Meng, S. Wu, L. Kuang, D. Huang, J. Lu, Multi-user detection for spatial modulation via structured approximate message passing. IEEE Commun. Lett. **20**(8), 1527–1530 (2016)

31. W. Wang, W. Zhang, Spatial modulation for uplink multi-user mmwave MIMO systems with hybrid structure. IEEE Trans. Commun. **68**(1), 177–190 (2019)
32. D. Perez-Adan, O. Fresnedo, J.P. Gonzalez-Coma, L. Castedo, Wideband user grouping for uplink multiuser mmwave MIMO systems with hybrid combining. IEEE Access **9**, 41360–41372 (2021)
33. Z. Huang, X. Cheng, A 3D non-stationary model for beyond 5G and 6G vehicle-to-vehicle mmWave massive MIMO channels. IEEE Trans. Intell. Transp. Systems, 1–17 (2021). https://doi.org/10.1109/TITS.2021.3077076

Chapter 6
Millimeter-Wave Index Modulation for Vehicular Downlink Transmission

Keywords mmWave · Massive multiple-input multiple-output · Precoded beamspace modulation · Hybrid block diagonalization · Matrix decomposition · Geometric mean decomposition · Fairness

6.1 Background

Based on the multi-user uplink access study, this chapter will showcase the solution to wideband mmWave vehicular downlink transmission. As mentioned in Chap. 5, IM has been recognized as a promising modulation technique for next-generation wireless communications owing to its energy efficiency and error performance advantages [1–5]. Similarly, the existing IM technologies for cmWave system cannot meet the stringent requirements of future vehicular network due to either the vulnerability under Doppler or incompatibility with 5G NR mmWave. To support high-speed and high-reliability vehicular communications, it is urgent to devise a new downlink mmWave IM scheme feasible for time-varying channels.

As a matter of fact, a similar problem has already been investigated in existing sub-6GHz systems with two representative IM solutions, namely the differential spatial modulation (DSM) [6] and the inter-carrier-interference cancellation assisted subcarrier index modulation (ICIC-SIM) [7]. Specifically, by utilizing the differential processing at the transmitter and the corresponding non-coherent detection schemes at the receiver in [6], DSM avoids the need for accurate CSI and therefore further reduces the computational complexity. In [7], the proposed ICIC-SIM system seizes both merits of index modulated orthogonal frequency division multiplexing (IM-OFDM) and inter-carrier interference (ICI) self-cancellation, which is simple yet effective and provides better system performance than IM-OFDM as well as ICI self-cancellation. Although both types demonstrate decent performance in time-varying channels, they are both unsuitable for mmWave systems [8]. First, both schemes fail to combine the multiple-input multiple-output (MIMO) and OFDM together, while MIMO OFDM is becoming the default configuration in mmWave systems [9–13]. Secondly, DSM and ICIC-SIM were originally designed for digital structures and are thus infeasible for mmWave systems that typically employ

X. Cheng et al., *mmWave Massive MIMO Vehicular Communications*, Wireless Networks, https://doi.org/10.1007/978-3-030-97508-1_6

a unique hybrid digital/analog structure. More importantly, for both DSM and ICIC-SIM, their anti-Doppler capabilities actually come at the expense of certain compromises. Specifically, for the former, a 3 dB performance loss is inevitable due to the differential design; while for the latter, a 50% spectrum loss would be incurred from the ICIC operation. In light of these intrinsic issues in space domain and subcarrier domain, we focus on the beamspace and aim to propose a new anti-Doppler beamspace IM design tailored for wideband mmWave mMIMO downlink systems.

To the best of our knowledge, reference [14] has proposed a beam index modulation (BIM) for mmWave MIMO communication scenario, where each communication node in the BIM system requires only one RF chain. Although BIM has demonstrated its advantages in terms of higher spectrum efficiency and reduced hardware complexity, its applicability is somewhat limited because it was designed for narrowband channels and only one RF chain is activated. In light of these intrinsic issues, we propose a new anti-Doppler IM design tailored for wideband mmWave mMIMO systems, termed wideband precoded beamspace modulation (wPBM), which not only generalizes BIM from the narrowband to wideband channels, but also further supports the activation of multiple beams instead of one beam.

The design of wPBM comprises three key elements. First, by selecting the optimal L beams with respect to all subcarriers, we generalize BIM developed in [14] from the hybrid narrowband to hybrid wideband systems and from activating one out of L beams to activating $R(R \geq 1)$ out of $L(L \geq R)$ beams. It is worth mentioning that the former extension is not straightforward because the analog precoder in hybrid OFDM systems is shared by all subcarriers, therefore wPBM has to be designed by jointly considering all subcarriers. As can be inferred from the name, wPBM includes both the analog part and the digital part. By naturally exploiting the analog part of the transceivers, the high-dimension high-correlation spatial channel is converted into the low-dimension and potentially low-correlation "virtual spatial channel" (VSC) encountered from the transmitter RF chains to receiver RF chains. Then, the digital precoder is designed based on the matrix decomposition scheme of VSC with geometric mean decomposition (GMD) [15]. Compared to singular value decomposition (SVD) algorithm, the GMD-based precoder design does not make tradeoffs between the throughput and the BER performance. With the help of GMD, the VSC is decomposed into multiple subchannels with identical signal-to-noise ratio (SNR), and hence the simple identical bit allocation can be used for all subchannels [16, 17]. Finally, a first-order Doppler compensator is tactfully positioned at the receiver, such that the modulated symbols will ultimately go through an equivalent time-invariant channel, whose parameters remain the same as the actual time-varying channel. Since both the frequency-selectivity and time-selectivity are addressed, the complicated block detection degenerates into simple subcarrier detection without any performance loss. It should be mentioned that the anti-Doppler capability of wPBM does not impose any cost, but comes from the effective exploitation of the hybrid system and the beamspace. Next, we will introduce the wPBM mmWave mMIMO system and channel models, wPBM transceiver design, the Doppler compensator for wPBM, simulations results and summary in detail. The numerical simulation can demonstrate the superiority over existing alternatives in doubly-selective channels.

6.2 Wideband Precoded Beamspace Modulation (wPBM)

6.2.1 System and Channel Models

As shown in Fig. 6.1, a wideband mmWave mMIMO system is considered in this chapter, where N_b-dimensional lens-antenna arrays with M_b RF chains and N_m-dimensional lens-antenna arrays with M_m RF chains are deployed at the Tx and Rx, respectively. The Tx communicates with the Rx via L data streams. For ease of representation, we consider a full-multiplexing case with $L = M_b = M_m$. It should be mentioned that such a condition can be relaxed into $L \leq \min\{M_b, M_m\}$, where those spare RF chains can be used for augmenting the diversity.

Similar to Chap. 5, we adopt a generic geometric model [18]. Specifically, given the configuration of ULAs, the tap-d channel sampled at instant-n $\mathbf{H}_d[n] \in \mathcal{C}^{N_m \times N_b}$ can be represented as

$$\mathbf{H}_d[n]=\sqrt{\frac{N_m N_b}{P}}\sum_{p=1}^{P}\alpha_p\delta(dT_s-\tau_p)\boldsymbol{a}_r(\theta_p)\boldsymbol{a}_t^*(\phi_p)e^{jw_p n}, d<D \tag{6.1}$$

where parameters consistent with Chap. 5. The corresponding inter-subcarrier channel response $\boldsymbol{H}_k[r] \in \mathcal{C}^{N_m \times N_b}(k, r \in [0, K-1])$ can be expressed as [19]

$$\boldsymbol{H}_k[r]=\frac{1}{K}\sum_{i=0}^{K-1}\sum_{d=0}^{N_c-1}\mathbf{H}_d[L_{cp}+i]e^{-j2\pi(rd+(k-r)i)/K}. \tag{6.2}$$

At subcarrier-k, let $\boldsymbol{s}_k = [s_{k,1}, s_{k,2}, \cdots, s_{k,L}]^T \in \mathcal{C}^{L\times 1}$ be the modulated symbol vector. Before being transmitted from antennas, $\boldsymbol{s}_k$ is first precoded by

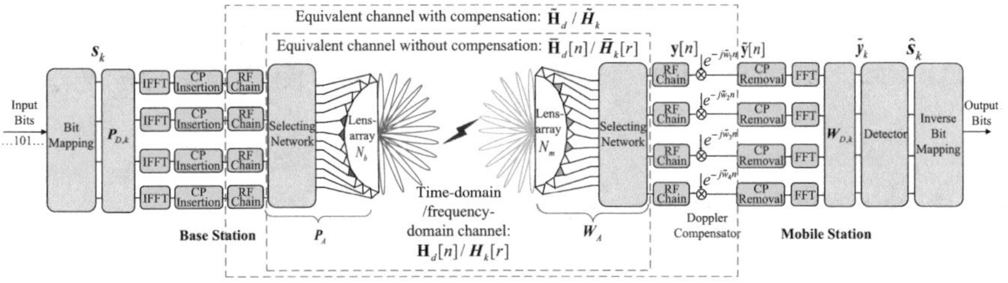

Fig. 6.1 An illustrative diagram of the mmWave mMIMO system with wPBM applied

$\boldsymbol{P}_{D,k}$, followed by IFFT operation and CP insertion. After being up-converted by RF chains, the resultant time-domain signal is processed by the analog precoder (selecting network plus lens-antenna array) $\boldsymbol{P}_A = [\boldsymbol{p}_T(t_1), \boldsymbol{p}_T(t_2), \cdots, \boldsymbol{p}_T(t_L)]$, with $\boldsymbol{p}_T(i)$ being the i-th DFT basis vector. Through channel propagation, the received signal on subcarrier-k is given by

$$\boldsymbol{r}_k = \boldsymbol{H}_k[k]\boldsymbol{P}_A\boldsymbol{P}_{D,k}\boldsymbol{s}_k + \sum_{r=0,r\neq k}^{K-1} \boldsymbol{H}_k[r]\boldsymbol{P}_A\boldsymbol{P}_{D,k}\boldsymbol{s}_r + \boldsymbol{w}_k \tag{6.3}$$

where $\boldsymbol{w}_k \sim \mathcal{CN}\left(\boldsymbol{0}, \sigma^2\boldsymbol{I}_{N_m}\right)$ is the Gaussian noise with σ^2 being the noise power.

Following the notation as the Tx-end, we use

$$\boldsymbol{W}_A = [\boldsymbol{f}_R(r_1), \boldsymbol{f}_R(r_2), \cdots, \boldsymbol{f}_R(r_L)]$$

to denote the Rx-end analog combiner. After CP removal and FFT operation, the frequency-domain signal will finally be combined by $\boldsymbol{W}_{D,k}$, then the signal to be processed becomes

$$\begin{aligned}\boldsymbol{y}_k &= \boldsymbol{W}_{D,k}^*\boldsymbol{W}_A^*\boldsymbol{H}_k[k]\boldsymbol{P}_A\boldsymbol{P}_{D,k}\boldsymbol{s}_k + \\ &\quad \sum_{r=0,r\neq k}^{K-1} \boldsymbol{W}_{D,k}^*\boldsymbol{W}_A^*\boldsymbol{H}_k[r]\boldsymbol{P}_A\boldsymbol{P}_{D,k}\boldsymbol{s}_r + \boldsymbol{\xi}_k \\ &= \boldsymbol{W}_{D,k}^*\bar{\boldsymbol{H}}_k[k]\boldsymbol{P}_{D,k}\boldsymbol{s}_k + \sum_{r=0,r\neq k}^{K-1} \boldsymbol{W}_{D,k}^*\bar{\boldsymbol{H}}_k[r]\boldsymbol{P}_{D,r}\boldsymbol{s}_r + \boldsymbol{\xi}_k\end{aligned} \tag{6.4}$$

where $\boldsymbol{\xi}_k = \boldsymbol{W}_{D,k}^*\boldsymbol{W}_A^*\boldsymbol{w}_k \sim \mathcal{CN}\left(\boldsymbol{0}, \sigma^2\boldsymbol{I}_L\right)$. By stacking all $\boldsymbol{y}_k$ into a long vector, we end up with the system I-O relationship as

$$\boldsymbol{y} = \boldsymbol{W}_D^* \underbrace{\begin{bmatrix} \bar{\boldsymbol{H}}_0[0] & \cdots & \bar{\boldsymbol{H}}_0[K-1] \\ \vdots & \ddots & \vdots \\ \bar{\boldsymbol{H}}_{K-1}[0] & \cdots & \bar{\boldsymbol{H}}_{K-1}[K-1] \end{bmatrix}}_{\bar{\boldsymbol{H}}} \boldsymbol{P}_D\boldsymbol{s} + \boldsymbol{\xi}. \tag{6.5}$$

with $\boldsymbol{W}_D = \text{diag}\{\boldsymbol{W}_{D,k}\}_{k=0}^{K-1}$, $\boldsymbol{P}_D = \text{diag}\{\boldsymbol{P}_{D,k}\}_{k=0}^{K-1}$, $\boldsymbol{s}$ and $\boldsymbol{\xi}$ being the concatenated modulated symbol vector $\{\boldsymbol{s}_k\}_{k=0}^{K-1}$ and noise vector $\{\boldsymbol{\xi}_k\}_{k=0}^{K-1}$, respectively.

6.2.2 wPBM Transceiver Design

In this subsection, we elaborate on the wPBM transceiver design, including bit mapping, Doppler compensation, detector and beam selection.

Table 6.1 An example of bit mapping with $b = 2$, $L = 2$, $R = 1$, $M = 2$

Input bits	Active beam index	Symbol	s_k
00	1	+1	$[+1, 0]^T$
01	1	−1	$[-1, 0]^T$
10	2	+1	$[0, +1]^T$
11	2	−1	$[0, -1]^T$

Let $b = b_1 + b_2$ be the number of input binary bits on each subcarrier. In wPBM, the first $b_1 = \lfloor \log_2(\mathbb{C}_L^R) \rfloor$ bits decide which R out of L beams will be activated, while the remaining $b_2 = R\log_2(M)$ bits will be mapped into R M-ary symbols according to a normalized constellation S, so the modulated vector $\boldsymbol{s}_k$ comprises constellation symbols as well as zero symbols. To gain a better intuition, an example of wPBM using binary phase shift keying (BPSK) is provided in Table 6.1. The spectral efficiency of wPBM is given by Basar et al. [5].

$$\eta = \frac{K}{K + L_{cp}} \left(\left\lfloor \log_2 \mathbb{C}_L^R \right\rfloor + R\log_2 M \right). \tag{6.6}$$

Let $\mathbf{y}[n] \in \mathcal{C}^{L\times 1}$ stand for the time-domain signal after RF chains (see Fig. 6.1). Typically, $\mathbf{y}[n]$ will be directly going to CP removal and FFT. However, in wPBM, an additional operation, i.e., a first-order compensation, will be applied to $\mathbf{y}[n]$. That is, the actual signal $\tilde{\mathbf{y}}[n] \in \mathcal{C}^{L\times 1}$ going to CP removal will be

$$\tilde{\mathbf{y}}[n] = \mathbf{diag}\left\{e^{-j\tilde{\omega}_l n}\right\}_{l=1}^{L} \mathbf{y}[n], 0 \le n \le K + L_{cp} - 1 \tag{6.7}$$

where $\tilde{\omega}_l$ is the associated Doppler shift with the selected beam-l. Apparently, such an operation adds negligible computational complexity.

Define $\bar{\mathbf{H}}_d[n] = \boldsymbol{W}_A^* \mathbf{H}_d[n] \boldsymbol{P}_A$. Let $\mathbf{x}[n] \in \mathcal{C}^{L\times 1}$ stand for the time-domain signal after IFFT and CP Insertion, then the received time-domain signal after compensation at instant-n accordingly becomes

$$\tilde{\mathbf{y}}[n] = \sum_{d=0}^{D-1} \mathbf{diag}\left\{e^{-j\tilde{\omega}_l n}\right\}_{l=1}^{L} \bar{\mathbf{H}}_d[n]\mathbf{x}[n-d] + \mathbf{v}[n] \tag{6.8}$$

where $\mathbf{v}[n] \sim \mathcal{CN}\left(\mathbf{0}, \sigma^2 \boldsymbol{I}_L\right)$. Then, $\tilde{\mathbf{y}}[n]$ in Eq. (6.8) can be further expressed as

$$\tilde{\mathbf{y}}[n] \approx \sum_{d=0}^{D-1} \tilde{\mathbf{H}}_d \mathbf{x}[n-d] + \mathbf{v}[n]. \tag{6.9}$$

Noticing that, unlike the original $\bar{\mathbf{H}}_d[n]$, $\tilde{\mathbf{H}}_d$ here proves to be irrespective of n. That means, the ICI has completely vanished during the transmission of $\boldsymbol{s}_k$, leaving the frequency response on subcarrier-k as

$$\tilde{\boldsymbol{H}}_k = \sum_{d=0}^{D-1} \tilde{\mathbf{H}}_d e^{-j\frac{2\pi k}{K}d}. \tag{6.10}$$

Therefore, the I-O relationship of wPBM can be written as

$$\tilde{\boldsymbol{y}} \approx \boldsymbol{W}_D^* \begin{bmatrix} \tilde{\boldsymbol{H}}_0 & \boldsymbol{O} & \cdots & \boldsymbol{O} \\ \boldsymbol{O} & \tilde{\boldsymbol{H}}_1 & \ddots & \vdots \\ \vdots & \ddots & \ddots & \boldsymbol{O} \\ \boldsymbol{O} & \cdots & \boldsymbol{O} & \tilde{\boldsymbol{H}}_{K-1} \end{bmatrix} \boldsymbol{P}_D \boldsymbol{s} + \boldsymbol{\xi}. \tag{6.11}$$

with $\boldsymbol{y}$ being the concatenated received symbol vector $\tilde{\boldsymbol{y}}_k$.

Without Doppler pre-compensation, the ML detector is expressed as

$$\left\{\hat{\boldsymbol{s}}_0, \cdots, \hat{\boldsymbol{s}}_{K-1}\right\} = \underset{\forall k, \boldsymbol{s}_k \in G_s}{\arg\min} \left\| \boldsymbol{y} - \boldsymbol{W}_D^* \bar{\boldsymbol{H}} \boldsymbol{P}_D [\boldsymbol{s}_0^T, \cdots, \boldsymbol{s}_{K-1}^T]^T \right\|^2 \tag{6.12}$$

where G_s contains all possible $\boldsymbol{s}_k$'s. Since the detection is performed across all subcarriers, the corresponding complexity in terms of multiplications is $O \sim (2^{bK})$, which tends to be prohibitive if K is large. Fortunately, with a simple first-order Doppler compensator, the block-based ML detector will degenerate into a low-complexity subcarrier-based one, i.e.,

$$\hat{\boldsymbol{s}}_k = \underset{\boldsymbol{s}_k \in G_s}{\arg\min} \left\| \tilde{\boldsymbol{y}}_k - \boldsymbol{W}_{D,k}^* \tilde{\boldsymbol{H}}_k \boldsymbol{P}_{D,k} \boldsymbol{s}_k \right\|^2. \tag{6.13}$$

The computational complexity is $O \sim (2^b K)$, same as in the static channels but with no performance loss, indicating that wPBM is a inherently anti-Doppler IM scheme.

6.2.3 Analog-Domain Processing

Note that those L beams to be shared by all subcarriers have been determined before the modulator and demodulator. Thus, we start with the analog-domain processing. Here, we set $\boldsymbol{W}_{D,k}^* = \boldsymbol{I}$ and

$$\boldsymbol{P}_{D,k} = \tilde{\boldsymbol{H}}_k^+ \left\| \tilde{\boldsymbol{H}}_k^+ \right\|_F^{-1}. \tag{6.14}$$

Obviously, $\boldsymbol{P}_{D,k}$ acts as the ZF pre-equalizer for $\tilde{\boldsymbol{H}}$. Define set $(\boldsymbol{r}, \boldsymbol{t}) = \{(r_1, t_1), \cdots, (r_L, t_L)\}$ with $\forall i, (r_i, t_i) \in \mathcal{P}$. Those L beam indices to be shared by all subcarriers are

Algorithm 4 Geometric mean decomposition (GMD)

1: **Input:** The equivalent digital-domain channel $\tilde{\boldsymbol{H}}_k$ at subcarrier-k;
2: **Initialization:** $SVD(\tilde{\boldsymbol{H}}_k) = \boldsymbol{U}_k\boldsymbol{\Sigma}_k\boldsymbol{V}_k$, the eigenvalues of $\boldsymbol{\Sigma}_k$ are $\delta_1, \delta_2, \cdots, \delta_{M_m}$, $(\delta_i \geq \delta_j, i > j)$. Then the geometric mean of the positive singular values is $\sigma = \sqrt[M_m]{\delta_1\delta_2\cdots\delta_{M_m}}$;
3: **For** $m = 1, \cdots, M_m - 1$, **do**
4: **If** $\delta_m \leq \sigma \leq \delta_{m+1}$ or $\delta_m \geq \sigma \geq \delta_{m+1}$, no change is made; **If not**, we exchange δ_m or δ_{m+1} with any element δ_j, $j > m + 1$, and rearrange the corresponding rows and columns in $\boldsymbol{U}_k$ and $\boldsymbol{V}_k$ with the permutation matrix $\boldsymbol{T}_m$.
5: **Transform** δ_m into geometric mean σ
$$\underbrace{\frac{1}{\sigma}\begin{bmatrix} c\delta_m & s\delta_{m+1} \\ -s\delta_{m+1} & c\delta_m \end{bmatrix}}_{\boldsymbol{\psi}_m^T}\begin{bmatrix} \delta_m & 0 \\ 0 & \delta_{m+1} \end{bmatrix}\underbrace{\begin{bmatrix} c & -s \\ s & c \end{bmatrix}}_{\boldsymbol{\Theta}_m} = \begin{bmatrix} \sigma & x \\ 0 & y \end{bmatrix}$$
Here, we focus on the transform of the relevant 2×2 submatrices. The rows and columns corresponding to the above two eigenvalues need to be converted. $c = 1$ and $s = 0$ if $\delta_m = \delta_{m+1} = \sigma$; $c = \sqrt{\frac{\sigma^2-\delta_2{}^2}{\delta_1{}^2-\delta_2{}^2}}$ and $s = \sqrt{1-c^2}$ if $\delta_m \neq \delta_{m+1}$;
6: **End For**
7: After the above operations, the diagonal matrix $\boldsymbol{\Sigma}_k$ has been transformed into an upper triangular matrix $\boldsymbol{R}_k$, compute
$$\boldsymbol{G}_k = \boldsymbol{U}_k \prod_{m=1}^{M_m-1}(\boldsymbol{T}_m\boldsymbol{\psi}_m);\ \boldsymbol{Q}_k = \boldsymbol{V}_k \prod_{m=1}^{M_m-1}(\boldsymbol{T}_m\boldsymbol{\Theta}_m)$$
8: **Output:** GMD matrix $\boldsymbol{G}_k$, $\boldsymbol{R}_k$, $\boldsymbol{Q}_k$ at subcarrier-k;

$$\left\{(\tilde{r}_1, \tilde{t}_1), \cdots, (\tilde{r}_L, \tilde{t}_L)\right\} = \arg\min_{(\boldsymbol{r},\boldsymbol{t})\subset\mathcal{P}} \max_{\forall k} Tr\left\{\{\tilde{\boldsymbol{H}}_k{}^*\tilde{\boldsymbol{H}}_k\}^{-1}\right\}. \tag{6.15}$$

Then the analog precoder and combiner are accordingly designed as

$$\boldsymbol{F}_A = [\boldsymbol{p}_T(\tilde{t}_1), \boldsymbol{p}_T(\tilde{t}_2), \cdots, \boldsymbol{p}_T(\tilde{t}_L)], \tag{6.16a}$$

$$\boldsymbol{W}_A = [\boldsymbol{f}_R(\tilde{r}_1), \boldsymbol{f}_R(\tilde{r}_2), \cdots, \boldsymbol{f}_R(\tilde{r}_L)]. \tag{6.16b}$$

6.2.4 Digital-Domain Processing

After analog-domain processing, we start to complete the digital-domain processing. In order to improve the BER performance, we introduce the GMD-based precoder to average the gain of each sub-channel. Assume that the channel $\tilde{\boldsymbol{H}}_k$ is known at the BS. Then we can design the corresponding precoding matrix and combing matrix by decomposing the channel $\tilde{\boldsymbol{H}}_k$.

For channel $\tilde{\boldsymbol{H}}_k$ with rank L at subcarrier k, we can factor it into

$$GMD(\tilde{\boldsymbol{H}}_k) = \boldsymbol{G}_k\boldsymbol{R}_k\boldsymbol{Q}_k^*, \tag{6.17}$$

where $\boldsymbol{G_k}$ and $\boldsymbol{Q_k}$ have orthogonal columns, and $\boldsymbol{R_k} \in \mathcal{C}^{L\times L}$ is a real upper triangular matrix with the non-zero diagonal elements all equal to the geometric mean of the positive singular values σ_k, as shown in Algorithm 4. Then we can obtain the corresponding analog precoder $\boldsymbol{P}_{D,k}$ and analog combiner $\boldsymbol{W}_{D,k}$ at subcarrier-k as follows

$$\boldsymbol{P}_{D,k} = \sqrt{L}\frac{\boldsymbol{Q}_k}{\left\|\boldsymbol{P}_A\boldsymbol{Q}_k\right\|_F}, \tag{6.18a}$$

$$\boldsymbol{W}_{A,k} = \boldsymbol{G}_k. \tag{6.18b}$$

then the received frequency-domain signal after compensation at subcarrier-k accordingly becomes

$$\begin{aligned}\tilde{\boldsymbol{y}}_k &= \boldsymbol{W}_{D,k}^*\tilde{\boldsymbol{H}}_k\boldsymbol{P}_{D,k}\boldsymbol{s}_k + \boldsymbol{\xi}_k \\ &= \boldsymbol{G}_k^*\tilde{\boldsymbol{H}}_k\boldsymbol{Q}_k\boldsymbol{s}_k + \boldsymbol{\xi}_k \\ &= \boldsymbol{R}_k\boldsymbol{s}_k + \boldsymbol{\xi}_k.\end{aligned} \tag{6.19}$$

So far we can get L sub-channels with exactly the same gain, which improves the BER performance of the system. We then rewrite Eq. (6.13) as

$$\hat{\boldsymbol{s}}_k = \underset{\boldsymbol{s}_k\in G_s}{\arg\min}\left\|\tilde{\boldsymbol{y}}_k - \boldsymbol{R}_k\boldsymbol{s}_k\right\|^2. \tag{6.20}$$

Because $\boldsymbol{R}_k$ is an upper triangular matrix, we can also use successive interference cancellation (SIC) scheme to reduce detection complexity. Assume

$$\boldsymbol{R}_k = \begin{bmatrix} R_{1,1} & R_{1,2} & \cdots & R_{1,L} \\ 0 & R_{2,2} & \ddots & \vdots \\ \vdots & \ddots & \ddots & R_{L-1,L} \\ 0 & \cdots & 0 & R_{L,L}\end{bmatrix}, \tag{6.21}$$

where all $R_{l,l}(l = 1, \cdots, L)$ are equal to each other. We can detect the received signal of each sub-channel on subcarrier-k one by one as follows

(1) Starting from the L-th subchannel, we can get

$$\hat{s}_{k,L} = \underset{s_{k,L}\in\{S,0\}}{\arg\min}\left\|y_{k,L} - R_{L,L}s_{k,L}\right\|^2.$$

Let $j = L - 1$, and go to step 2;

(2) Using the detected $\hat{s}_{k,i}(i = j+1, \cdots, L)$ for interference cancellation, we can get

$$\hat{y}_{k,j} = y_{k,j} - \sum_{i=j+1}^{L} R_{i,L}\hat{s}_{k,i}$$

(3) Using $\hat{y}_{k,j}$, we further obtain

$$\hat{s}_{k,j} = \underset{s_{k,j}\in\{S,0\}}{\arg\min} \left\| \hat{y}_{k,j} - R_{L,j}s_{k,j} \right\|^2.$$

(4) Repeating steps (2) and (3) above until detecting all streams.

6.3 Extension to Multi-User Setup

6.3.1 Design Motivation

With judicious transceiver design, the resultant mmWave massive mMIMO can not only enhance the link quality by harnessing power gains but also facilitate spatial multiplexing by refining beam patterns [20]. Owing to these prominent merits, one mmWave mMIMO BS is expected to serve multiple UEs simultaneously. The resultant wideband multi-user (wMU) mmWave mMIMO systems pave the way towards massive connection and ultra-fast speed for the next-generation cellular. Thus, we aim to extend the downlink single-user wPBM system to wMU scenarios to enable next-generation mmWave vehicular cellular.

As mentioned in Chap. 5, for the wMU mmWave mMIMO uplink system, multi-user signal detection for different multiple access methods at the BS is the main research point. While in the wMU mmWave mMIMO downlink transmission, the mMU case encounters three significant challenges in transceiver design. The first stems from the unique hybrid structure, which imposes more constraints on the transceiver design [21]. The second issue still associates with the unique structure, in which the per-subcarrier processing with OFDM applied is no longer independent due to a shared analog beamformer. Last but not least, unlike the point-to-point scenarios, the wMU scenarios have to jointly consider the user-specific signal quality variation and the multi-user interference (MUI) [22]. For the non-cooperative downlink system, the locations of each user are independent of each other, and the CSI of the remaining users cannot be obtained. Therefore, MUI is difficult to achieve on the mobile terminal. However, the BS has the ability to obtain the CSI of the users it serves, so the MUI operation can be carried out on the BS side, that is, the BS can precode the transmitted data to simplify the signal receiving operation on the mobile side.

6.3.2 Overall Strategy

In practice, a so-termed hybrid block diagonalization (HBD) framework is commonly adopted for transceiver design [23]. Essentially, HBD requires the transmitter to eliminate the MUI with the help of CSI, such that independent detection can be applied individually at the UE end. The most representative wideband HBD schemes are proposed in [24–26]. Yuan et al. [24] and Liu and Zhu [25] are essentially a greedy extension of its point-to-point ancestors, inheriting their limitation and lacking a proper MUI management, thereby failing to offer a performance guarantee for general wMU mmWave mMIMO systems. Gao et al. [26] follows the popular HBD-based paradigm but mitigates the potential ad-hoc or empirical nature, and then proposed a systematic transceiver solution for wMU mmWave mMIMO based on the criterion of MI maximization. However, the mentioned wMU mmWave mMIMO systems do not combine the superiority of beamspace IM in terms of the BER performance and power efficiency. Thus, wMU mmWave mMIMO with wPBM is not only a natural extension of single user model, but also the realistic requirements of next generation mobile communication.

Against this background of HBD, we transform the complicated wMU hybrid expansion processing of wPBM into a well-tractable two-stage analog-domain plus digital-domain processing. First, based on the beam selection criterion of minimizing the mean square error function, we end up with sub-optimal yet low-complexity analog-domain processing. Secondly, we complete the digital-domain processing to facilitate MUI-free reception and individual BER improvements. The entire design does not impose any requirement on the angular resolution or the channel sparsity, making it a more general and appealing HBD solution to downlink multi-user transmission.

6.3.3 System and Channel Models

In this section, We extend the downlink wPBM system from wideband single-user to wMU over doubly-selective channels. We first introduce the system and channel models. Based on the derived I-O relationship, we then formulate the corresponding transceiver design for wMU mmWave mMIMO.

A downlink wMU mmWave mMIMO system is considered in this section, with an illustrative diagram shown in Fig. 6.2. A BS with N_b-dimensional lens-antenna arrays and M_b RF chains at the Tx serves U mobile users at the Rx. Each UE is equipped with N_m-dimensional lens-antenna arrays and M_m RF chains. We assume that the BS communicates to each UE via N_s data streams, satisfying $UN_s \leq M_b \leq N_b$ and $N_s \leq M_m \leq N_m$. For ease of representation, we assume $N_s = M_m$ and $M_b = UN_s$.

Similar to Chap. 5, the channel parameters from the BS to different UEs are distributed independently. For the channel in time domain from the BS to the UE-u,

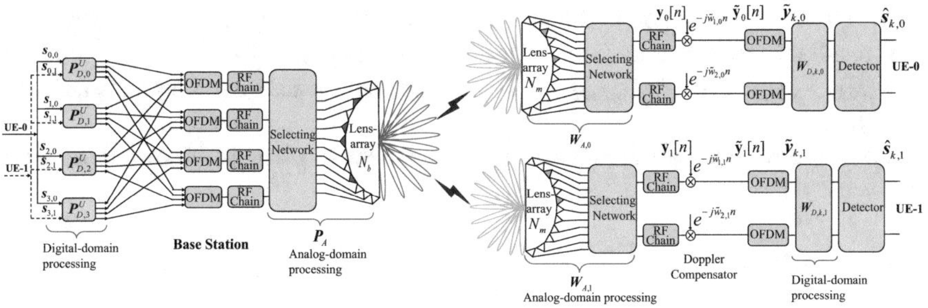

Fig. 6.2 An illustrative wMU mmWave mMIMO system model with $K = 4, U = 2, M_b = 4, M_m = 2$

the tap-d channel at time instant-n $\boldsymbol{H}_{d,u}[n] \in \mathcal{C}^{N_m \times N_b}$ can be written as:

$$\sqrt{\frac{N_b N_m}{P}} \sum_{p=1}^{P} \alpha_{p,u} p_{rc}(dT_s - \tau_{p,u}) \boldsymbol{a}_r(\theta_{p,u}) \boldsymbol{a}_t^*(\phi_{p,u}) e^{j w_{p,u} n}, \tag{6.22}$$

where the parameters are consistent with Chap. 5. Then, under the commonly used half-wavelength spaced ULAs, we have $\boldsymbol{a}_t(\phi_{p,u}) = \boldsymbol{f}_{N_b}(\pi \sin(\phi_{p,u}))$ and $\boldsymbol{a}_r(\theta_{p,u}) = \boldsymbol{f}_{N_m}(\pi \sin(\theta_{p,u}))$. Given the time-domain channel response, the corresponding inter-subcarrier channel response $\boldsymbol{H}_{k,u}[r] \in \mathcal{C}^{N_m \times N_b}(k, r \in [0, K-1])$ can be expressed as

$$\boldsymbol{H}_{k,u}[r] = \frac{1}{K} \sum_{i=0}^{K-1} \sum_{d=0}^{N_c-1} \boldsymbol{H}_{d,u}[L_{cp} + i] e^{-j2\pi(rd+(k-r)i)/K}. \tag{6.23}$$

Apparently, when $w_{p,u}$=0, $\mathbf{H}_{d,u}[n]$ turns out to be time-invariant. Denote the static version of $\mathbf{H}_{d,u}[n]$ as $\mathbf{H}_{d,u}$, then in this case, for UE-u, all the inter-subcarrier channels vanish except for leaving

$$\boldsymbol{H}_{k,u}[k] = \boldsymbol{H}_{k,u} = \sum_{d=0}^{N_c-1} \mathbf{H}_{d,u} e^{-j\frac{2\pi k}{K}d}. \tag{6.24}$$

At the BS, $\boldsymbol{s}_k^U = \left[\mathbf{s}_{k,0}^T, \mathbf{s}_{k,1}^T, \cdots \mathbf{s}_{k,U-1}^T\right]^T$, is a $UM_m \times 1$ signal vector transmitted to U UEs. $\mathbf{s}_{k,u}$ will be sent to UE-u with its symbols selected from a Gaussian constellation. $\boldsymbol{s}_k^U$ is first precoded by BS-end digital precoder

$$\boldsymbol{P}_{D,k}^U = \left[\boldsymbol{P}_{D,k,1}, \boldsymbol{P}_{D,k,2}, \ldots, \boldsymbol{P}_{D,k,U}\right] \in \mathcal{C}^{M_b \times UM_m} \tag{6.25}$$

followed by M_b K-point IFFTs. Then, a CP of length L_{cp} is added to the resultant time-domain signal before applying the RF precoder $\boldsymbol{P}_A = [\boldsymbol{f}_T(t_1), \boldsymbol{f}_T(t_2), \cdots, \boldsymbol{f}_T(t_{M_b})] \in \mathcal{C}^{N_b \times M_b}$ (selecting network plus lens-antenna array).

After the BS-end hybrid precoding, the signal captured by UE-u on subcarrier k is given by

$$\boldsymbol{r}_{k,u} = \boldsymbol{H}_{k,u}[k]\boldsymbol{P}_A \boldsymbol{P}_{D,k}^U \boldsymbol{s}_k^U + \sum_{r=0, r\neq k}^{K-1} \boldsymbol{H}_{k,u}[r]\boldsymbol{P}_A \boldsymbol{P}_{D,r}^U \boldsymbol{s}_r^U + \boldsymbol{w}_{k,u}, \tag{6.26}$$

where $\boldsymbol{w}_{k,u} \sim CN\left(\boldsymbol{0}, \sigma^2 \boldsymbol{I}_{N_m}\right)$ is the Gaussian noise with σ^2 being the noise power at UE-u.

Following the notation as the Tx-end, we use

$$\boldsymbol{W}_{A,u} = [\boldsymbol{f}_R(r_{1,u}), \boldsymbol{f}_R(r_{2,u}), \cdots, \boldsymbol{f}_R(r_{M_m,u})]$$

to denote analog combiner at UE-u. By further removing CP and implementing M_m K-point FFT operation, the frequency-domain signal will finally be combined by $\boldsymbol{W}_{D,k,u}$, resulting in

$$\begin{aligned}\boldsymbol{y}_{k,u} &= \boldsymbol{W}_{D,k,u}^* \boldsymbol{W}_{A,u}^* \boldsymbol{H}_{k,u}[k]\boldsymbol{P}_A \boldsymbol{P}_{D,k}^U \boldsymbol{s}_k^U + \\ &\quad \sum_{r=0, r\neq k}^{K-1} \boldsymbol{W}_{D,k,u}^* \boldsymbol{W}_{A,u}^* \boldsymbol{H}_{k,u}[r]\boldsymbol{P}_A \boldsymbol{P}_{D,r}^U \boldsymbol{s}_r^U + \boldsymbol{\xi}_{k,u} \\ &= \boldsymbol{W}_{D,k,u}^* \bar{\boldsymbol{H}}_{k,u}[k]\boldsymbol{P}_{D,k}^U \boldsymbol{s}_k^U + \\ &\quad \sum_{r=0, r\neq k}^{K-1} \boldsymbol{W}_{D,k,u}^* \bar{\boldsymbol{H}}_{k,u}[r]\boldsymbol{P}_{D,r}^U \boldsymbol{s}_r^U + \boldsymbol{\xi}_{k,u}\end{aligned} \tag{6.27}$$

where $\boldsymbol{\xi}_{k,u} = \boldsymbol{W}_{D,k,u}^* \boldsymbol{W}_{A,u}^* \boldsymbol{w}_k \sim CN\left(\boldsymbol{0}, \sigma^2 \boldsymbol{I}_L\right)$.

Similar to [26], we use the block diagonalization (BD) algorithm to eliminate inter-user interference. Thus, the designed $\boldsymbol{P}_{D,r}^U$, which will be introduced later, can guarantee

$$\bar{\boldsymbol{H}}_{k,u}[r]\boldsymbol{P}_{D,r}^U = \left[\underbrace{\boldsymbol{O}, \ldots, \boldsymbol{O}}_{u-1}, \bar{\boldsymbol{H}}_{k,u}[r]\boldsymbol{P}_{D,r,u}, \underbrace{\boldsymbol{O}, \ldots, \boldsymbol{O}}_{U-u}\right] \tag{6.28}$$

Let $\boldsymbol{s}_u^U = \left[\mathbf{s}_{0,u}^T, \mathbf{s}_{1,u}^T, \cdots \mathbf{s}_{K-1,u}^T\right]^T$. The wideband multi-user I-O relationship between the BS and UE-u can be expressed as

$$\mathbf{y}_u^U = \boldsymbol{W}_{D,u}^* \underbrace{\begin{bmatrix} \bar{\boldsymbol{H}}_{0,u}[0] & \cdots & \bar{\boldsymbol{H}}_{0,u}[K-1] \\ \vdots & \ddots & \vdots \\ \bar{\boldsymbol{H}}_{K-1,u}[0] & \cdots & \bar{\boldsymbol{H}}_{K-1,u}[K-1] \end{bmatrix}}_{\bar{\boldsymbol{H}}_u} \boldsymbol{P}_{D,u}^U \mathbf{s}_u^U + \boldsymbol{\xi}_u^U \tag{6.29}$$

with $\boldsymbol{W}_{D,u} = \text{diag}\{\boldsymbol{W}_{D,k,u}\}_{k=0}^{K-1}$, $\boldsymbol{P}_{D,u}^U = \text{diag}\{\boldsymbol{P}_{D,k,u}^U\}_{k=0}^{K-1}$, $\mathbf{s}_u^U$ and $\boldsymbol{\xi}_u^U$ being the concatenated modulated symbol vector $\{\mathbf{s}_{k,u}\}_{k=0}^{K-1}$ and noise vector $\{\boldsymbol{\xi}_{k,u}\}_{k=0}^{K-1}$, respectively.

6.3.4 wPBM wMU Transceiver Design

In this section, we elaborate on the wideband multi-user transceiver design, including bit mapping, doppler compensation, detector, analog-domain processing and digital-domain processing.

For each UE-u, let $b = b_1 + b_2$ be the number of input binary bits on each subcarrier. The first $b_1 = \left\lfloor \log_2(\mathbb{C}_{M_m}^R) \right\rfloor$ bits decide which R out of M_m beams will be activated, while the remaining $b_2 = R\log_2(M)$ bits will be mapped into R M-ary symbols according to a normalized constellation S, so the modulated vector $\mathbf{s}_{k,u}$ comprises constellation symbols as well as zero symbols. The spectral efficiency of wideband multi-user system is given by Basar et al. [5]

$$\eta = \frac{UK}{K + L_{cp}} \left(\left\lfloor \log_2 \mathbb{C}_{M_m}^R \right\rfloor + R\log_2 M \right). \tag{6.30}$$

Let $\mathbf{y}_u[n] \in \mathcal{C}^{M_m \times 1}$ stand for the time-domain signal after RF chains (see Fig. 6.2). After compensation, the actual signal $\tilde{\mathbf{y}}_u[n] \in \mathcal{C}^{M_m \times 1}$ going to CP removal will be

$$\tilde{\mathbf{y}}_u[n] = \mathbf{diag}\left\{ e^{-j\tilde{\omega}_{l,u}n} \right\}_{l=1}^{L} \mathbf{y}_u[n], 0 \le n \le K + L_{cp} - 1 \tag{6.31}$$

where $\tilde{\omega}_{l,u}$ is the associated Doppler shift with the selected beam-l for UE-u.

Define $\bar{\mathbf{H}}_{d,u}[n] = \boldsymbol{W}_{A,u}^* \mathbf{H}_{d,u}[n] \boldsymbol{P}_A$. The received time-domain signal after compensation at instant-n accordingly becomes

$$\tilde{\mathbf{y}}_u[n] = \sum_{d=0}^{D-1} \mathbf{diag}\left\{ e^{-j\tilde{\omega}_{l,u}n} \right\}_{l=1}^{L} \bar{\mathbf{H}}_{d,u}[n] \mathbf{x}[n-d] + \mathbf{v}_u[n] \tag{6.32}$$

where $\mathbf{v}_u[n] \sim CN\left(\mathbf{0}, \sigma^2 \boldsymbol{I}_L\right)$. Then, $\tilde{\mathbf{y}}_u[n]$ in Eq. (6.32) can be further expressed as

$$\tilde{\mathbf{y}}_u[n] \approx \sum_{d=0}^{D-1} \tilde{\mathbf{H}}_{d,u}\mathbf{x}[n-d] + \mathbf{v}_u[n]. \tag{6.33}$$

leaving the frequency response on subcarrier-k as

$$\tilde{\boldsymbol{H}}_{k,u} = \sum_{d=0}^{D-1} \tilde{\mathbf{H}}_{d,u} e^{-j\frac{2\pi k}{K}d}. \tag{6.34}$$

Therefore,the wideband multi-user I-O relationship between the BS and UE-u in Eq. (6.29) can be rewritten as

$$\boldsymbol{y}_u^U \approx \boldsymbol{W}_{D,u}^* \begin{bmatrix} \tilde{\boldsymbol{H}}_{0,u} & \boldsymbol{O} & \cdots & \boldsymbol{O} \\ \boldsymbol{O} & \tilde{\boldsymbol{H}}_{1,u} & \ddots & \vdots \\ \vdots & \ddots & \ddots & \boldsymbol{O} \\ \boldsymbol{O} & \cdots & \boldsymbol{O} & \tilde{\boldsymbol{H}}_{K-1,u} \end{bmatrix} \boldsymbol{P}_{D,u}^U \boldsymbol{s}_u^U + \boldsymbol{\xi}_u^U. \tag{6.35}$$

If no Doppler pre-compensation is applied, the ML detector is

$$\left\{\hat{\boldsymbol{s}}_{0,u}, \cdots, \hat{\boldsymbol{s}}_{K-1,u}\right\} = \underset{\forall k, \boldsymbol{s}_{k,u} \in G_s}{\arg\min} \left\| \boldsymbol{y}_u^U - \boldsymbol{W}_{D,u}^* \bar{\boldsymbol{H}}_u \boldsymbol{P}_{D,u}^U [\boldsymbol{s}_{0,u}^T, \cdots, \boldsymbol{s}_{K-1,u}^T]^T \right\|^2 \tag{6.36}$$

where G_s contains all possible $\boldsymbol{s}_{k,u}$'s. Same to the single-user case, we apply the first-order Doppler compensation, giving rise to

$$\hat{\boldsymbol{s}}_{k,u} = \underset{\boldsymbol{s}_{k,u} \in G_s}{\arg\min} \left\| \tilde{\boldsymbol{y}}_{k,u} - \boldsymbol{W}_{D,k,u}^* \tilde{\boldsymbol{H}}_{k,u} \boldsymbol{P}_{D,k,u} \boldsymbol{s}_{k,u} \right\|^2. \tag{6.37}$$

Stacking all $\{\boldsymbol{y}_{k,u}\}_{k=0}^{K-1}$, the multi-user I-O relationship accordingly becomes

$$\boldsymbol{y}_k^U = \boldsymbol{W}_{D,k}^{U\,*} \underbrace{\begin{bmatrix} \tilde{\boldsymbol{H}}_{k,1} \\ \tilde{\boldsymbol{H}}_{k,2} \\ \vdots \\ \tilde{\boldsymbol{H}}_{k,U} \end{bmatrix}}_{\tilde{\boldsymbol{H}}_k^U} \boldsymbol{P}_{D,k}^U \boldsymbol{s}_k^U + \boldsymbol{\xi}_k^U. \tag{6.38}$$

with $\boldsymbol{W}_{D,k} = \mathrm{diag}\{\boldsymbol{W}_{D,k,u}\}_{u=0}^{U-1}$, $\boldsymbol{P}_{D,k}^U = \mathrm{diag}\{\boldsymbol{P}_{D,k,u}^U\}_{u=0}^{U-1}$, $\boldsymbol{s}_k^U$ and $\boldsymbol{\xi}_k^U$ being the concatenated modulated symbol vector $\{\boldsymbol{s}_{k,u}\}_{u=0}^{U-1}$ and noise vector $\{\boldsymbol{\xi}_{k,u}\}_{u=0}^{U-1}$, respectively.

Define set $(\boldsymbol{r}_u, \boldsymbol{t}_u) = \left\{(r_{1,u}, t_{1,u}), \cdots, (r_{M_m,u}, t_{M_m,u})\right\}$ with $\forall i$, $(r_{i,u}, t_{i,i}) \in \mathcal{P}_u$. Those UM_m beam indices to be shared by all subcarriers are

$$\begin{aligned}&\left\{(\tilde{r}_{1,0}, \tilde{t}_{1,0}), \cdots, (\tilde{r}_{l,u}, \tilde{t}_{l,u}), \cdots, (\tilde{r}_{M_m,U-1}, \tilde{t}_{M_m,U-1})\right\}\\ &= \underset{(\boldsymbol{r}_u,\boldsymbol{t}_u)\subset\mathcal{P}_u}{\arg\min}\ \max_{\forall k} Tr\left\{\{\tilde{\boldsymbol{H}}_k^{U*}\tilde{\boldsymbol{H}}_k^U\}^{-1}\right\}.\end{aligned} \tag{6.39}$$

Then the analog precoder and combiner are accordingly designed as

$$\boldsymbol{P}_A = [\boldsymbol{p}_T(\tilde{t}_{1,0}), \cdots, \boldsymbol{p}_T(\tilde{t}_{M_m,0}), \cdots, \boldsymbol{p}_T(\tilde{t}_{1,U-1}), \cdots, \boldsymbol{p}_T(\tilde{t}_{M_m,U-1})] \tag{6.40a}$$

$$\boldsymbol{W}_{A,u} = [\boldsymbol{f}_R(\tilde{r}_{1,u}), \boldsymbol{f}_R(\tilde{r}_{2,u}), \cdots, \boldsymbol{f}_R(\tilde{r}_{M_m,u})]. \tag{6.40b}$$

After analog-domain processing, we start to complete the digital-domain processing targeting MUI removal and further reducing the system's the wideband multi-user' BER while improving the system channel capacity. Without loss of generality, we take UE-u on subcarrier-k as an example. To highlight the twofold functionality of digital-domain processing, we specially decompose the associated digital precoder as

$$\boldsymbol{P}_{D,k,u} = \boldsymbol{P}_{D,k,u,1}\boldsymbol{P}_{D,k,u,2} \tag{6.41}$$

where $\boldsymbol{P}_{D,k,u,1} \in \mathcal{C}^{M_b\times M_m}$is for MUI removal, $\boldsymbol{P}_{D,k,u,2} \in \mathcal{C}^{M_m\times M_m}$ is for individual BER performance improvement.

(1) First-Step Digital-Domain Processing In order to get an MUI-free equivalent digital channel that satisfied Eq. (6.28), we set $\widehat{\boldsymbol{H}}_k[u]$ as

$$\widehat{\boldsymbol{H}}_{k,u} = \left[\tilde{\boldsymbol{H}}_{k,0}^T, \cdots, \tilde{\boldsymbol{H}}_{k,u-1}^T, \tilde{\boldsymbol{H}}_{k,u+1}^T, \cdots, \tilde{\boldsymbol{H}}_{k,U-1}^T\right]^T \tag{6.42}$$

then we know $\boldsymbol{P}_{D,k,u,1}$ must lie in the null-space of $\widehat{\boldsymbol{H}}_k[u]$.

For $M_b \geq UM_m$, we consider both the interference null space and the signal space. Based on the theory of subspace projection and vector space of matrices [26], the interference null space $\boldsymbol{F}_{k,u}^{null}$ and signal space $\boldsymbol{F}_{k,u}^{signal}$ can be written as

$$\boldsymbol{F}_{k,u}^{null} = \boldsymbol{I} - \widehat{\boldsymbol{H}}_{k,u}^*(\widehat{\boldsymbol{H}}_{k,u}\widehat{\boldsymbol{H}}_{k,u}^*)^{-1}\widehat{\boldsymbol{H}}_{k,u} \tag{6.43a}$$

$$\boldsymbol{F}_{k,u}^{signal} = \tilde{\boldsymbol{H}}_{k,u}^*(\tilde{\boldsymbol{H}}_{k,u}\tilde{\boldsymbol{H}}_{k,u}^*)^{-1}\tilde{\boldsymbol{H}}_{k,u}. \tag{6.43b}$$

By performing $SVD(\boldsymbol{F}_{k,u}^{signal}\boldsymbol{F}_{k,u}^{null}) = \widehat{\boldsymbol{U}}_{k,u}\widehat{\boldsymbol{\Sigma}}_{k,u}\widehat{\boldsymbol{V}}_{k,u}^*$, $\boldsymbol{P}_{D,k,u,1}$ can be set as $\widehat{\boldsymbol{V}}_{k,u}[:,1:M_m]$.

(2) Second-Step Digital-Domain Processing Thanks to $\boldsymbol{P}_{D,k,u,1}$, along with the analog precoders, UE-u ends up with an MUI-free EDC on subcarrier-k as follows:

$$\tilde{\boldsymbol{H}}_{eff,k,u} = \tilde{\boldsymbol{H}}_{k,u}\boldsymbol{P}_{D,k,u,1}, \tag{6.44}$$

The received signal can, therefore, be expressed as

$$\tilde{\boldsymbol{y}}_{k,u} = \tilde{\boldsymbol{H}}_{eff,k,u}\boldsymbol{s}_{k,u} + \boldsymbol{\xi}_{k,u} \tag{6.45}$$

Similar to single user, we use GMD algorithm to complete the second-step digital-domain processing, then we can obtain

$$GMD(\tilde{\boldsymbol{H}}_{eff,k,u}) = \boldsymbol{G}_{eff,k,u}\boldsymbol{R}_{eff,k,u}\boldsymbol{Q}^{*}_{eff,k,u} \tag{6.46}$$

$\boldsymbol{Q}_{eff,k,u}$ and $\boldsymbol{G}_{eff,k,u}$ are the digital precoders for $\tilde{\boldsymbol{H}}_{eff,k,u}$ at the transmitter and receiver, respectively. By setting $\boldsymbol{P}_{D,k,u,2} = \boldsymbol{Q}_{eff,k,u}$ and taking transmit power constraint into account, the digital precoders are finally determined as

$$\boldsymbol{P}_{D,k,u} = \sqrt{M_m}\frac{\boldsymbol{P}_{D,k,u,1}\boldsymbol{Q}_{eff,k,u}}{\left\|\boldsymbol{P}_{A}\boldsymbol{P}_{D,k,u,1}\boldsymbol{Q}_{eff,k,u}\right\|_F} \tag{6.47a}$$

$$\boldsymbol{W}_{A,k,u} = \boldsymbol{G}_{eff,k,u}. \tag{6.47b}$$

Up to this point, we have accomplish the entire downlink transceiver design for wideband multi-user mmWave mMIMO.

6.4 Simulations

In this section, we consider single-user scene and multi-user scenarios, and compare wPBM with the conventional spatial multiplexing (CSM), the maximum beamforming (MBF) and wPBM without doppler compensation (DC) in terms of BER performance. CSM uses the M_m selected beams without IM for transmission, while MBF uses the strongest beam for transmission. The system related parameters are set as: $N_b = 32, 64$, $N_m = 32$ with $L = 2$. The channel related parameters are set as: $K = 256$, $L_{cp} = D = 64$, $P = 10$ and $f_c = 60$ GHz. $h(\cdot)$ is set as the raised-cosine filter with a roll-off factor $\beta = 0.8$. The SNR is defined as $\frac{S}{N_0} = \frac{RU}{\sigma^2}$.

6.4.1 BER in Doubly-Selective Channels

Before investigating doubly-selective channels, we first make comparisons in frequency-selective time-invariant (FSTI) channels. Specifically, in Fig. 6.3, we fix $\eta = 2.4$ bps/Hz with $N_b = 32, 64$. wPBM sets $R = 1$ with 4-quadrature amplitude modulation (4-QAM). CSM adopts BPSK and 4-QAM. MBF adopts

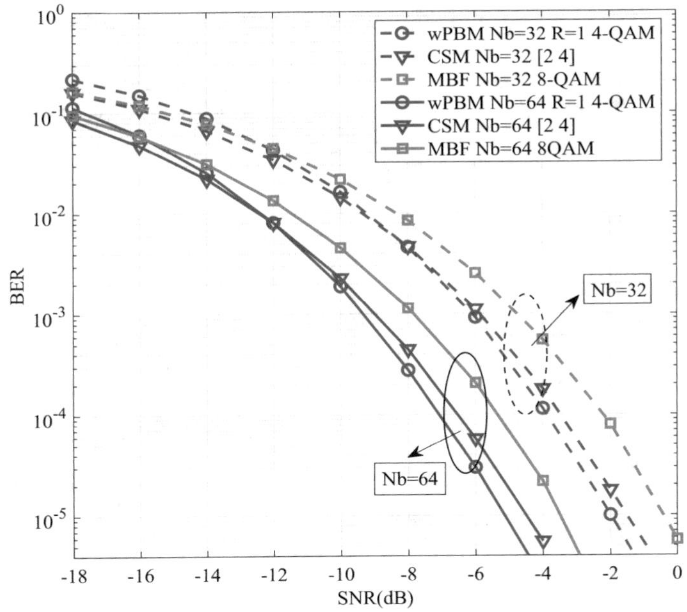

Fig. 6.3 BER comparisons among wPBM, CSM, and MBF in FSTI channels ($U = 1, N_b = 32, 64$)

8-QAM modulation. Figure 6.3 illustrates that at low SNR, wPBM is inferior to its counterparts, because the index bits are prone to strong noises thus could be more likely detected by mistake. However, once SNR becomes modestly large, the performance of wPBM soon exceeds its counterparts, with a coding gain of 2 and 0.7 dB over MBF and CSM. This advantages are owing to 33% ratio of index bits, respectively, which are more robust in high SNR region.

Following the above system configuration, we then compare the LMMSE-based and SIC-based detectors with N_b equal to 32 and 64 in Fig. 6.4. The modulation parameters in Fig. 6.4 keeps identical to those in Fig. 6.3. Simulations show that the LMMSE-based detector achieves a coding gain of about 2 dB over the SIC-based detector at BER = 10^{-6}. This BER advantage is obtained at the cost of increasing detection complexity. Although yielding a lower detection complexity, this one is unfriendly in guiding beam selection as the LMMSE-based detector does.

After demonstrating the advantage of wPBM in FSTI channels, we then introduce the Doppler effects and verify whether this advantage still holds in frequency-selective time-variant (FSTV) channels. In Fig. 6.5, apart from comparing with wPBM without compensation in FSTV channels, we also compare with wPBM in the corresponding FSTI channels, whose performance is set as the ideal benchmark. The modulation parameters in Fig. 6.5 keep identical to those in Fig. 6.3 with $N_b = 64$. For FSTV channels, three different levels of normalized maximum Doppler

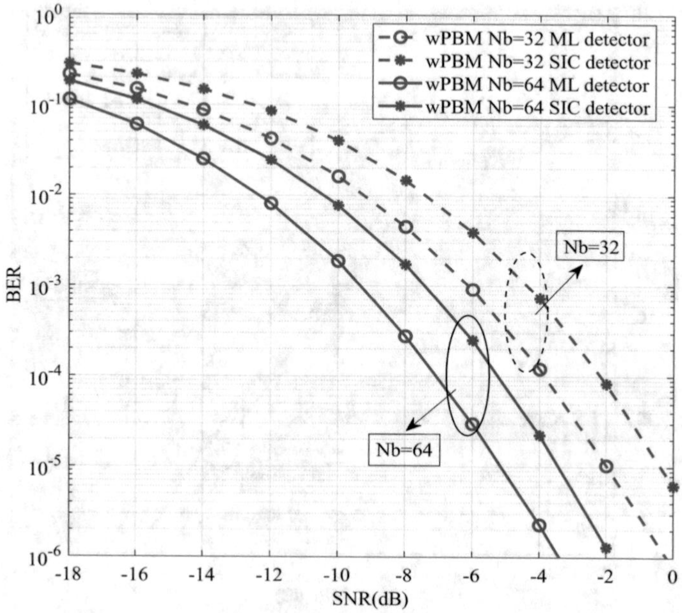

Fig. 6.4 BER comparisons of wPBM with different detector ($U = 1$, $N_b = 32, 64$)

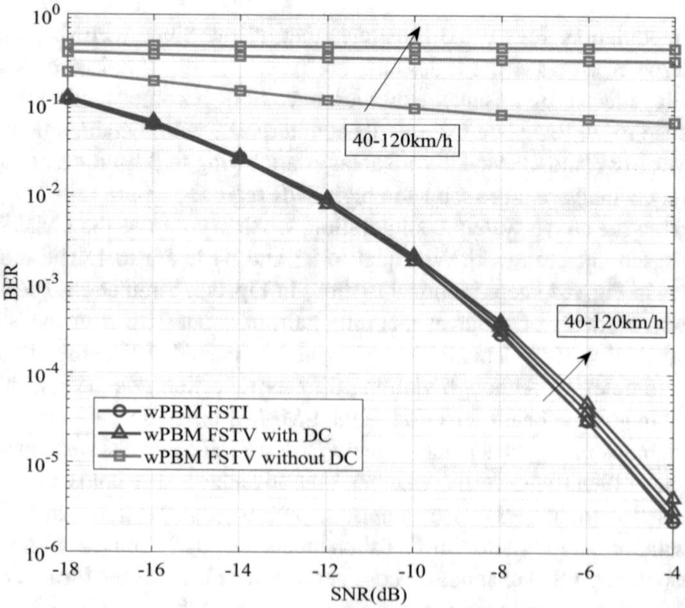

Fig. 6.5 BER comparisons among wPBM, CSM, and MBF in FSTV channels ($U = 1$, $N_b = 64$)

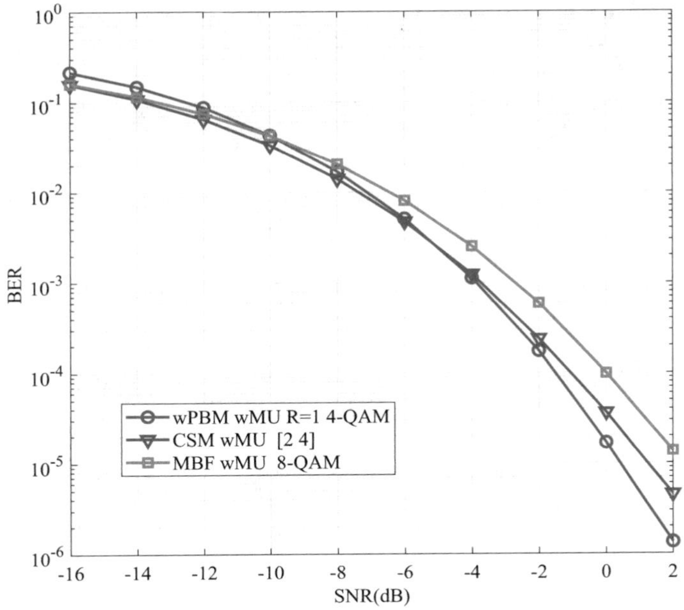

Fig. 6.6 BER comparisons among wPBM, CSM, and MBF in FSTI channels ($U = 3$, $N_b = 64$)

shifts are considered: 0.005, 0.01, and 0.015, corresponding to $v_m = 40$, 80, and 120 km/h, respectively. As can be seen from Fig. 6.5, wPBM performs dramatically better than wPBM without DC. In Fig. 6.5, the gap between the compensated wPBM and the ideal benchmark is minimal at relatively low SNR; while in the high SNR region, there exists a small BER gap as v_m increases. This is because the anti-Doppler nature of wPBM is not at the expense of a certain sacrifice, but coming from the exploitation of beamspace and the careful design of the Doppler pre-compensator.

We then set $U = 3$ and test the extended wPBM wMU scheme over doubly-selective channels with SE equal to 2.4 bps/Hz in Figs. 6.6 and 6.7. The modulation parameters keep identical to those in Fig. 6.3 with $N_b = 64$. Similar to sing-user wGBM, we observe that wGBM wMU is inferior to CSM at low SNR, while at high SNR, wMU wGBM outperforms CSM. Then, we test whether the wMU BER advantage still holds in FSTV channels in Fig. 6.7. The performance of wGBM wMU in FSTI channels is set as the ideal benchmark. Here, we also consider three different levels of maximum Doppler shift: 0.005, 0.01, and 0.015, corresponding to $v_m = 40$, 80, and 120 km/h, respectively. We observe that wGBM wMU also performs well at three levels of mobility.

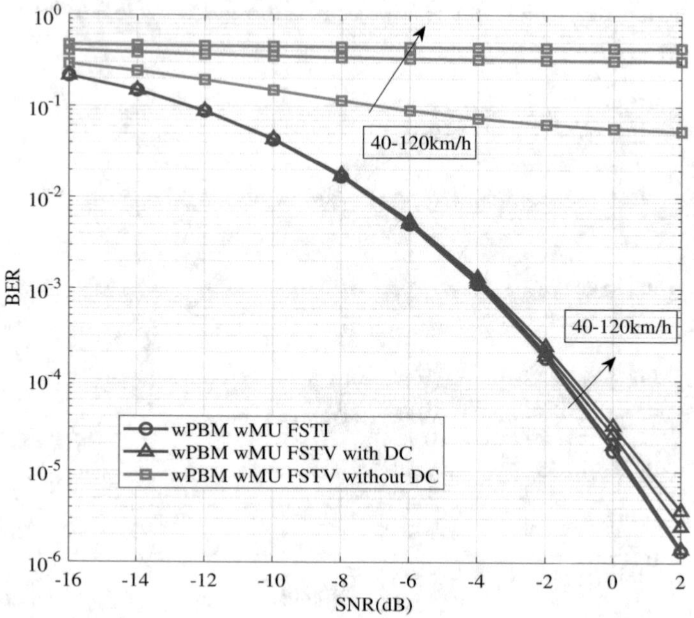

Fig. 6.7 BER comparisons among wGBM, CSM, and MBF in FSTV channels ($U = 3$, $N_b = 64$)

6.4.2 BER in Low Vehicular Traffic Density (VTD) Slow-Mobility Non-stationary Channels

We further test the proposed wGBM wMU scheme using a low vehicular traffic density (VTD) slow-mobility non-stationary beyond 5G and 6G vehicular mmWave mMIMO channels [27] in Fig. 6.8. The modulation parameters keeps identical to those in Fig. 6.6. We observe that wGBM wMU also outperforms others. At high SNR, wGBM wMU achieves a coding gain of 1 and 2.2 dB over CSM and MBF, respectively. The maintenance of such an advantage is because the proposed scheme does not impose much constraint on the channel itself. Therefore, varying the AoAs/AoDs distribution or beam direction does not affect its feasibility.

6.5 Discussions and Summary

In this section, we tailored an index modulation, namely, wPBM for mmWave vehicular downlink access. wPBM can effectively combat the time selectivity and remain compatible with hybrid mmWave wideband systems. Towards a multi-user extension, the HBD framework ensures high flexibility at the BS end and low detection complexity at the user end. Besides, applying GMD within a well-

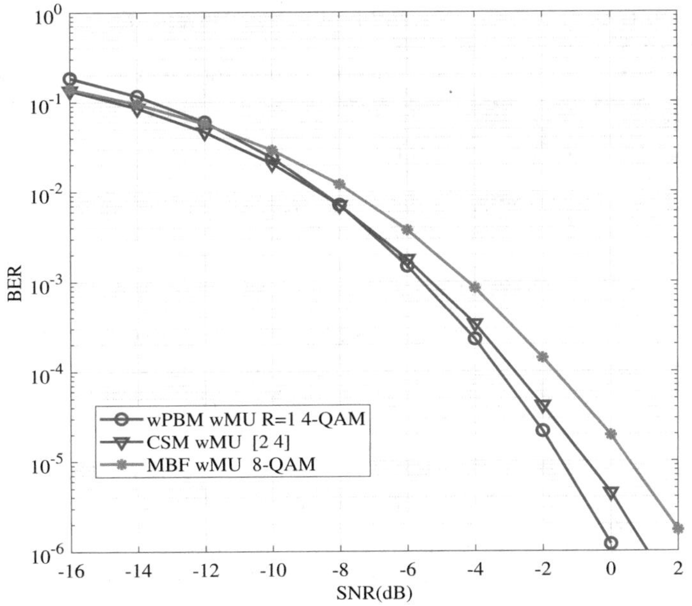

Fig. 6.8 BER comparisons among wGBM, CSM, and MBF in low VTD slow-mobility Non-stationary beyond 5G and 6G vehicular mmWave mMIMO channels ($U = 3$, $N_b = 64$)

designed beamspace channel has effectively guaranteed fairness in terms of the BER. Owing to these prominent merits, wPBM would be a promising candidate for downlink transmissions.

Based on wPBM, one can explore two more directions to augment the downlink transmission. First, we can incorporate power allocation and adaptive modulation into the BS-end design. This will endow the BS with more power in determining how to best distribute resources per the user's channel quality. Secondly, we can extend from single-cell to multi-cell scenarios such that the entire index modulation design needs particular dedication to manage multiple sources of interference.

References

1. X. Cheng, M. Zhang, M. Wen, L. Yang, Index modulation for 5G: Striving to do more with less. IEEE Wirel. Commun. **25**(2), 126–132 (2018)
2. M. Wen, X. Cheng, L. Yang, *Index Modulation for 5G Wireless Communications* (Springer, Cham, 2017)
3. R. Mesleh, H. Haas, S. Sinanovic, C.W. Ahn, S. Yun, Spatial modulation. IEEE Trans. Veh. Technol. **57**(4), 2228–2241 (2008)
4. M. Di Renzo, H. Haas, A. Ghrayeb, S. Sugiura, L. Hanzo, Spatial modulation for generalized MIMO: Challenges, opportunities, and implementation. Proc. IEEE **102**(1), 56–103 (2014)
5. E. Basar, U. Aygolu, E. Panayirci, H.V. Poor, Orthogonal frequency division multiplexing with index modulation. IEEE Trans. Signal Process. **61**(22), 5536–5549 (2013)

6. Y. Bian, X. Cheng, M. Wen, L. Yang, H.V. Poor, B. Jiao, Differential spatial modulation. IEEE Trans. Veh. Technol. **64**(7), 3262–3268 (2015)
7. Y. Li, M. Wen, X. Cheng, L. Yang, Index modulated OFDM with ICI self-cancellation for V2X communications, in *2016 International Conference on Computing, Networking and Communications (ICNC)* (2016), pp. 1–5
8. S. Han, I. Chih-Lin, Z. Xu, C. Rowell, Large-scale antenna systems with hybrid analog and digital beamforming for millimeter wave 5G. IEEE Commun. Mag. **53**(1), 186–194 (2015)
9. Y. Lin, Hybrid MIMO-OFDM beamforming for wideband mmwave channels without instantaneous feedback. IEEE Trans. Signal Process. **66**(19), 5142–5151 (2018)
10. J. Zhang, Y. Huang, J. Wang, L. Yang, Hybrid precoding for wideband millimeter-wave systems with finite resolution phase shifters. IEEE Trans. Veh. Technol. **67**(11), 11285–11290 (2018)
11. F. Sohrabi, W. Yu, Hybrid analog and digital beamforming for mmwave OFDM large-scale antenna arrays. IEEE J. Sel. Areas Commun. **35**(7), 1432–1443 (2017)
12. W. Shen, X. Bu, X. Gao, C. Xing, L. Hanzo, Beamspace precoding and beam selection for wideband millimeter-wave MIMO relying on lens antenna arrays. IEEE Trans. Signal Process. **67**(24), 6301–6313 (2019)
13. J. Li, S. Gao, X. Cheng, L. Yang, Hybrid precoded spatial modulation (hPSM) for mmwave massive MIMO systems over frequency-selective channels. IEEE Wirel. Commun. Lett. **9**(6), 839–842 (2020)
14. Y. Ding, V. Fusco, A. Shitvov, Y. Xiao, H. Li, Beam index modulation wireless communication with analog beamforming. IEEE Trans. Veh. Technol. **67**(7), 6340–6354 (2018)
15. Y. Jiang, W. Hager, J. Li, The geometric mean decomposition. Linear Algebra Appl. **396**, 373–384 (2005)
16. Y. Jiang, J. Li, W. Hager, Joint transceiver design for MIMO communications using geometric mean decomposition. IEEE Trans. Signal Process. **53**(10), 3791–3803 (2005)
17. C.-E. Chen, Y.-C. Tsai, C.-H. Yang, An iterative geometric mean decomposition algorithm for MIMO communications systems. IEEE Trans. Wirel. Commun. **14**(1), 343–352 (2015)
18. S. Gao, X. Cheng, L. Yang, Estimating doubly-selective channels for hybrid mmwave massive MIMO systems: A doubly-sparse approach. IEEE Trans. Wirel. Commun. **19**(9), 5703–5715 (2020)
19. L. Rugini, P. Banelli, G. Leus, OFDM communications over time-varying channels, in *Wireless Communications Over Rapidly Time-Varying Channels*, ed. by F. Hlawatsch, G. Matz (Academic Press, New York, 2011)
20. Z. Pi, F. Khan, An introduction to millimeter-wave mobile broadband systems. IEEE Commun. Mag. **49**(6), 101–107 (2011)
21. S. Gao, X. Cheng, L. Yang, Spatial multiplexing with limited RF chains: Generalized beamspace modulation (GBM) for mmwave massive MIMO. IEEE J. Sel. Areas Commun. **37**(9), 2029–2039 (2019)
22. L. Liang, W. Xu, X. Dong, Low-complexity hybrid precoding in massive multiuser MIMO systems. IEEE Wirel. Commun. Lett. **3**(6), 653–656 (2014)
23. W. Ni, X. Dong, Hybrid block diagonalization for massive multiuser MIMO systems. Trans. Commun. **64**(1), 201–211 (2016)
24. H. Yuan, J. An, N. Yang, K. Yang, T. Duong, Low complexity hybrid precoding for multiuser millimeter wave systems over frequency selective channels. IEEE Trans. Veh. Technol. **68**(1), 983–987 (2019)
25. B. Liu, H. Zhu, Rotman lens-based two-tier hybrid beamforming for wideband mmwave MIMO-OFDM system with beam squint. EURASIP J. Wirel. Commun. Netw. **267**, 1–13 (2018)
26. S. Gao, X. Cheng, L. Yang, Mutual information maximizing wideband multi-user (wMU) mmwave massive MIMO. IEEE Trans. Commun. **69**(5), 3067–3078 (2021)
27. Z. Huang, X. Cheng, A 3-D non-stationary model for beyond 5G and 6G vehicle-to-vehicle mmwave massive mimo channels. IEEE Trans. Intell. Transp. Syst., 1–17 (2021). https://doi.org/10.1109/TITS.2021.3077076

Index

X. Cheng et al., *mmWave Massive MIMO Vehicular Communications*, Wireless Networks, https://doi.org/10.1007/978-3-030-97508-1

T

V

W

Printed in the United States
by Baker & Taylor Publisher Services